Process Development Division 2017

Core Programming Area at the 2017 AIChE Spring Meeting and 13th Global Congress on Process Safety

San Antonio, Texas, USA
26 – 30 March 2017

ISBN: 978-1-5108-4146-8

Printed from e-media with permission by:

Curran Associates, Inc.
57 Morehouse Lane
Red Hook, NY 12571

Some format issues inherent in the e-media version may also appear in this print version.

Printed by Curran Associates, Inc. (2017)

For permission requests, please contact AIChE
at the address below.

AIChE
120 Wall Street, FL 23
New York, NY 10005-4020

Phone: (800) 242-4363
Fax: (203) 775-5177

www.aiche.org

Additional copies of this publication are available from:

Curran Associates, Inc.
57 Morehouse Lane
Red Hook, NY 12571 USA
Phone: 845-758-0400
Fax: 845-758-2633
Email: curran@proceedings.com
Web: www.proceedings.com

TABLE OF CONTENTS

482030 AGILE Award Keynote Address

Abstract:

Factors in Selecting AGILE Honoree

1. Overall sustained excellent industry leadership in businesses that employ chemical engineers

2. Recognized for efforts in innovation and creating strategies that ensure success

3. Value of the keynote to the meeting attendees and the members of the Institute Schedule

Event	Time	Location
Continental Breakfast	7:00AM - 8:00AM	Henry Gonzalez Convention Center, Lila Cockrell Theatre
AGILE Keynote Address	8:00AM - 9:15AM	Henry Gonzalez Convention Center, Lila Cockrell Theatre

The 2017 AIChE Spring Meeting and 13th GCPS AGILE Keynote Address will be presented by AGILE award recipient John Y. Televantos, Partner, Arsenal Capital Partners.

John Televantos is a Partner at Arsenal Capital Partners, a middle market private equity firm focused on the specialty industrial, healthcare and specialty financial service sectors. Before joining Arsenal in 2006, Dr. Televantos was President of Aqualon Company, owned by Hercules, Inc.; President and Founder of Helios Chemical Company; and Chief Executive Officer for Foamex International.

Dr. Televantos began his career with Union Carbide in 1977 in silicones research. There he developed two key processes for the manufacture of Methyl Chloride and Siloxane polymers that were commercialized. These processes not only improved efficiency and reduced cost but also dramatically reduced waste. He moved through a variety of management positions there and later at ARCO Chemical, where he became vice president of strategy and development.

At ARCO John co-invented and led a team that developed and commercialized a breakthrough technology in polyol manufacture (a key raw material for polyurethanes) eliminating a whole process step along with waste and emissions, while at the same time opening the door to the creation of new higher-performance polyurethanes. The technology has since been licensed by most of the major global polyol manufacturers with significant waste reduction worldwide.

At Arsenal Capital, Dr. Televantos serves as co-head of the specialty industrial practice. During his tenure, he has been part of several acquisitions and successful exits in the chemicals and materials sectors. In addition to serving on the boards of several Arsenal-owned companies, he is lead director on the board of Balchem Corp.

479724 Fluidization: Bubble, Bubble, Toil and Trouble

Abstract:

A quick dip into the open literature concerning fluidization quickly leaves one with a headache. The open literature is rife with empirical correlations relating pressure drop to fluid flow, material bed height, material particle diameter, and density differences. The same literature is also replete with "rules of thumb" for the onset of fluidization and for the various stages of fluidization. Undoubtedly, companies employing fluidized bed technology have similar documentation describing their own correlations and their own "rules of thumb".

In this presentation, we derive the pertinent relations for describing fluidization using Dimensional Analysis. We also discuss scaling fluidized processes from laboratory-sized equipment to commercial-sized equipment.

481004 Managing Heat Loss in Pilot Plant Operations to Simulate Full Scale Operation

Abstract:

Pilot plant operations present unique challenges regarding heat loss to the environment. The volume of chemicals and sizes of the equipment typically result in a high surface to volume ratio that results in excessive heat losses. Understanding and managing these heat losses (or gains for cold services) is critical to successful scale-up as the losses or gains from the atmosphere can be a significant part of the plant's energy balance.

For example, a 2" distillation column with no heat tracing is likely to lose heat to the atmosphere, resulting in cooling on each tray and giving a false result when interpreting the separations data and trying to fit it to process models. Heat tracing the column on the other hand, if not done correctly, will result in hot spots in the column and similarly poor matches to process models. If the thermophysical properties are adjusted to match the columns in these conditions, the result could be a poor model that is used for scale-up to the full commercial process. The result can easily be a commercial column that doesn't perform as expected.

This presentation will explore similar past experiences with heat transfer to the atmosphere in small scale equipment and some of the more successful methods to understand or mitigate these effects.

480842 Solids Handling Challenges during Process Scale-up

Abstract:

Solids handling operations are sometimes not given enough emphasis during planning and design stages. This is often based on the perception that solids will relatively easily move from bins, silos, hoppers, transfer chutes and feeders. However, this is not the case many times, especially when you are scaling up operations. During the pilot-plant stage, due to the low production volumes, it is possible to rely on manual intervention for assisting solids handling. However, when the pilot plant successfully demonstrates the concept and it is time for scale-up, one cannot rely on manual intervention. The solids handling operations need to perform reliably, on their own. Many studies have shown that when handling solids, the time required to start-up a plant can be significantly more than the planned time, due to challenges involved in solids handling.

This presentation will review different solids handling operations at a typical production plant, the challenges associated with reliably handling the solids and what can be done to avoid solids flow issues.

479208 Laboratory-Scale Integrated ARP Filter Test

Abstract:

The Savannah River Site (SRS) is currently treating radioactive liquid waste with the Actinide Removal Process (ARP) and the Modular Caustic Side Solvent Extraction Unit (MCU). Recently, the low filter flux through the ARP of approximately 5 gallons per minute has limited the rate at which radioactive liquid waste can be treated. Salt Batch 6 had a lower processing rate and required frequent filter cleaning. There is a desire to understand the causes of the low filter flux and to increase ARP/MCU throughput.

One plausible cause of the low filter flux is plugging of the primary (i.e., crossflow) or secondary filters. Much of the testing that was conducted to design and develop the filtration process in ARP focused on one aspect of the process (e.g., filtration, washing, chemical cleaning) rather than the entire process.

This work attempted to simulate the entire ARP process, including multiple batches (5), washing, chemical cleaning, and blending the feed with heels and recycle streams. The objective of the tests was to determine whether one of these processes is causing excessive fouling of the crossflow or secondary filter. The authors conducted the tests with feed solutions containing 6.6 M sodium Salt Batch 6 simulant supernate with no MST.

This presentation will describe the test design and the test results.

Reduced Energy Requirements for Direct Chlorination of Ethylene to Ethylene Dichloride

Kyle Self, Chemetry, Moss Landing, CA, USA

Michael Weiss, Chemetry, Moss Landing, CA, USA

Ryan Gilliam, Chemetry, Moss Landing, CA, USA

Marshall Lee, TechnipFMC, Boston, MA, USA

Vincent Welch, TechnipFMC, Boston, MA, USA

Abstract

The eShuttle™ process eliminates the need for chlorine gas generation in the synthesis of ethylene dichloride. The new process uses the same feedstocks (brine and ethylene) to produce ethylene dichloride, caustic soda, and hydrogen. The main benefit of the new process is reduced electrical power consumption. By replacing the conventional chlor-alkali anode reaction with a transition metal oxidation reaction, the electrolysis cell operates at substantially lower voltage and without the safety issues associated with chlorine gas generation, compression and storage. Over the past 36 months, the eShuttle™ process has been successfully implemented and tested at Chemetry's pilot scale facility in Moss Landing, CA. The overall process economics and fundamental theory of technology are also discussed.

Introduction

The existing vinyls value chain:

The chlor-alkali reaction is the most widely used electrochemical production process in the chemicals industry. In this reaction, chlorine and caustic are generated via the electrolysis of an aqueous sodium chloride brine:

$$2NaCl + 2H_2O \rightarrow Cl_2 + 2NaOH + H_2$$

Because the Gibbs Free Energy for the above reaction is positive, driving the reaction forward requires external work in the form of electricity. Thus, the process is best described as two independent half-cell reactions, each shown below with their voltages vs. the standard hydrogen electrode (SHE):

Anode Reaction:	$2Cl^- \rightarrow Cl_{2(g)} + 2e^-$	(-1.36V vs. SHE)
Cathode Reaction:	$2H_2O + 2e^- \rightarrow H_{2(g)} + 2OH^-$	(+0.83V vs. SHE)

Combining the voltages of the two reactions in an electrolysis cell results in a theoretical minimum voltage of 2.2V. However, commercial membrane cells operate at voltages above

3.0V, where the additional voltage is required to compensate for losses such as membrane resistance, solution resistances, and required over-potentials.

A major consumer of chlorine is the polyvinyl chloride (PVC) value chain, which represents about 34% of the downstream market.[1] To produce PVC, the next step is the manufacture of ethylene dichloride (EDC) which is made through two commercial pathways. The first pathway is direct chlorination, where chlorine reacts with ethylene to produce EDC:

$$C_2H_4 + Cl_2 \rightarrow C_2H_4Cl_2$$

Oxychlorination, the second pathway, uses ethylene, oxygen, and hydrochloric acid (HCl), the latter typically formed as a by-product of other processes (see below):

$$C_2H_4 + 2HCl + \frac{1}{2}O_2 \rightarrow C_2H_4Cl_2 + H_2O$$

EDC formed from both reactions is then thermally cracked to produce vinyl chloride monomer (VCM):

$$C_2H_4Cl_2 \rightarrow C_2H_3Cl + HCl$$

Finally, the VCM is then polymerized to form PVC for end use. The HCl byproduct from EDC cracking is typically recycled to the oxychlorination unit to produce more EDC. As a result, all the chloride atoms contained PVC were originally sourced as chlorine in the chlor-alkali process. The entire vinyls chain is shown below in Figure 1:

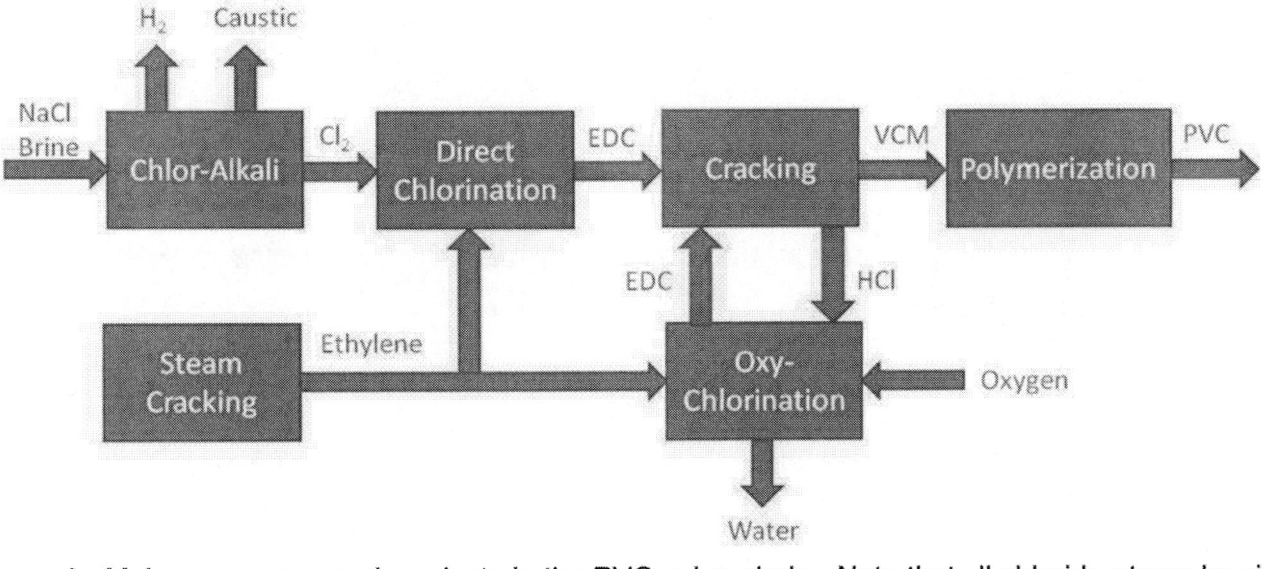

Figure 1: Major processes and products in the PVC value chain. Note that all chloride atoms leaving in PVC were originally sourced via the chlor-alkali process.

The opportunity:

The ability to replace reactions requiring chlorine gas with safer, lower-energy alternatives represents a great opportunity. Chlorine production from the chlor-alkali process

[1] See for example, https://www.ihs.com/products/chlorine-sodium-chemical-economics-handbook.html

is electrically intensive because two electrons must be transported over a 3V potential difference for every molecule of chlorine generated. As a result, chlor-alkali power demands often represent a significant fraction of the electrical grid consumption in developed countries. For example, chlor-alkali energy usage (just over 4 GW) represents approximately 1% of total US grid consumption[2]. In addition to the recognized direct costs associated with energy demand, chlorine production has significant hidden costs associated with safety risks in storage and transport.[3] Finally, because of the high power demand, conventional chlorine production results in a high carbon footprint due to associated power plant CO_2 emissions. As a result, replacement of conventional chlorine production methods remains an important environmental, safety, and economic objective.

The eShuttle™ Process

eShuttle™ Process Chemistry:

Chemetry's eShuttle™ process uses a recirculating, aqueous stream of copper to shuttle chloride ions from NaCl to EDC. This eliminates chlorine generation from the PVC value chain by replacing the combined chlor-alkali and direct chlorination processes as shown below in Figure 2:

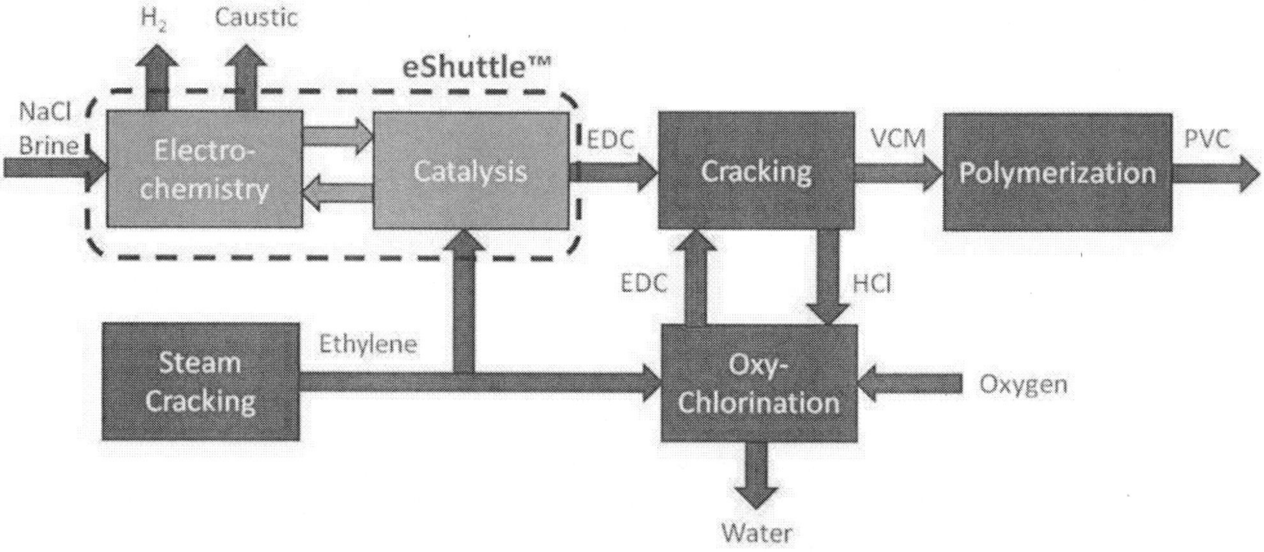

Figure 2: The eShuttle™ process replaces the chlor-alkali and direct chlorination processes with a single process that operates at lower energies and avoids chlorine as an intermediate.

Comparing Figure 1 with Figure 2 illustrates that the eShuttle™ process and the conventional chlor-alkali/direct-chlorination processes have the same feedstock inputs (NaCl

---

[2] Chlor-alkali energy consumption based on internal TechnipFMC calculations. US energy grid consumption of 447GW from https://en.wikipedia.org/wiki/List_of_countries_by_electricity_consumption

[3] See for example the US Center for Disease Control (CDC) report: "Top Five Chemicals Resulting in Injuries from Acute Chemical Incidents — Hazardous Substances Emergency Events Surveillance, Nine States, 1999–2008" available at https://www.cdc.gov/mmwr/preview/mmwrhtml/ss6402a6.htm.

brine and ethylene), product outputs (hydrogen, caustic and EDC) and, consequently, the same overall reaction:

$$2NaCl + 2H_2O + C_2H_4 \rightarrow C_2H_4Cl_2 + 2NaOH + H_2$$

The difference is that the chlorine gas intermediate has been eliminated in the eShuttle™ process.

As shown below in Figure 3, the eShuttle™ technology uses a recirculating stream of aqueous copper chloride. In the electrochemistry portion of the process, (shown on the left below), the cathode reaction remains the same as the chlor-alkali process. However, the anode reaction has now been replaced with the oxidation of cuprous chloride to cupric chloride. A second membrane, for anion exchange, prevents the copper ions from migrating into the brine stream. The design of the middle brine compartment, the development of the anion exchange membrane, and the engineering of the anode itself all represent major advances in reducing the cell voltage further improving the power advantages of the process.

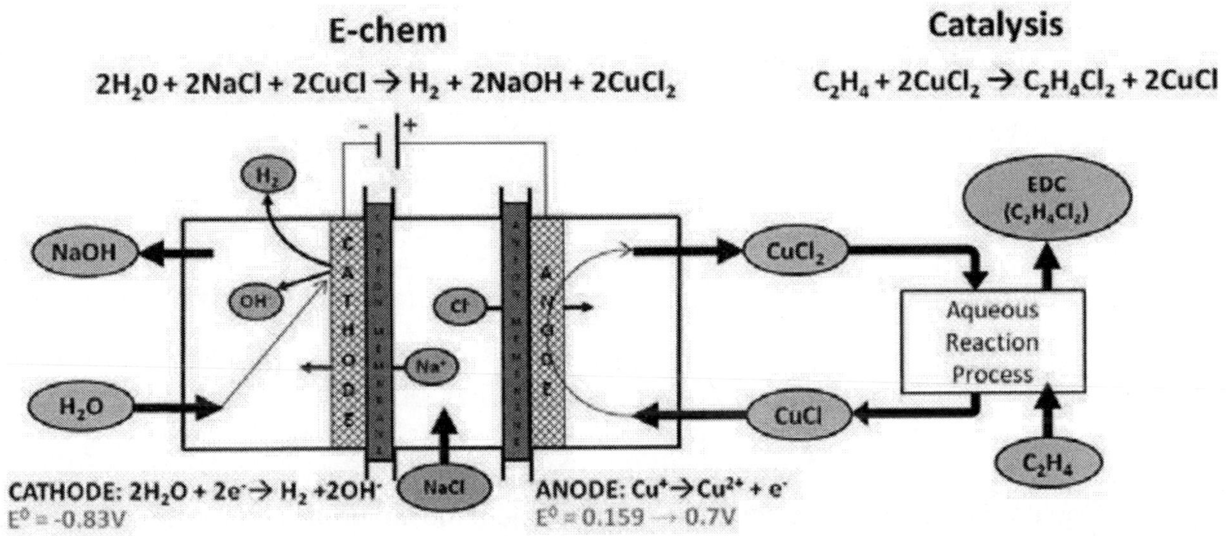

Figure 3: The eShuttle™ process circulates a continuous stream of aqueous copper chloride salts in a closed loop transporting Cl between the electrochemical cell and the process reactor.

The cupric chloride stream is then directed to a reaction system where it is contacted with ethylene to produce EDC under pressure and at elevated temperature. The optimal temperature and pressure is determined by balancing a combination of several variables including selectivity, space-time yield (STY) and capital costs.

A major breakthrough of the eShuttle™ process is the replacement of the chlor-alkali anode reaction, substituting chlorine generation (-1.36V vs. SHE) with a metal redox couple. The current version of eShuttle™ utilizes copper(I) oxidation as the anode reaction:

Anode Reaction: $Cu^{+}_{(aq)} \rightarrow Cu^{2+}_{(aq)} + e^{-}$ (-0.16V vs. SHE)

Thus, the theoretical minimum for this cell reaction should be about 1V. However, the copper speciation in the eShuttle™ process differs from the dilute ions assumed in the standard reaction above. That is the half cell reaction for the eShuttle™ process is more correctly written as the oxidation of cuprous chloride[4]:

Anode Reaction: $CuCl + Cl^- \rightarrow CuCl_2 + e^-$ (~ -0.70V vs. SHE)

Thus, the minimum voltage is actually 1.5V. Factoring in the losses due to membrane resistances and over potential, Chemetry operates its cells at about 2.4V, based on economic optimizations for commercial scale. As shown below in Figure 4, compared to state-of-the-art membrane electrolysis, the net result is a savings of 400kWh/tonne savings (on a caustic basis) – a major step change in energy performance.

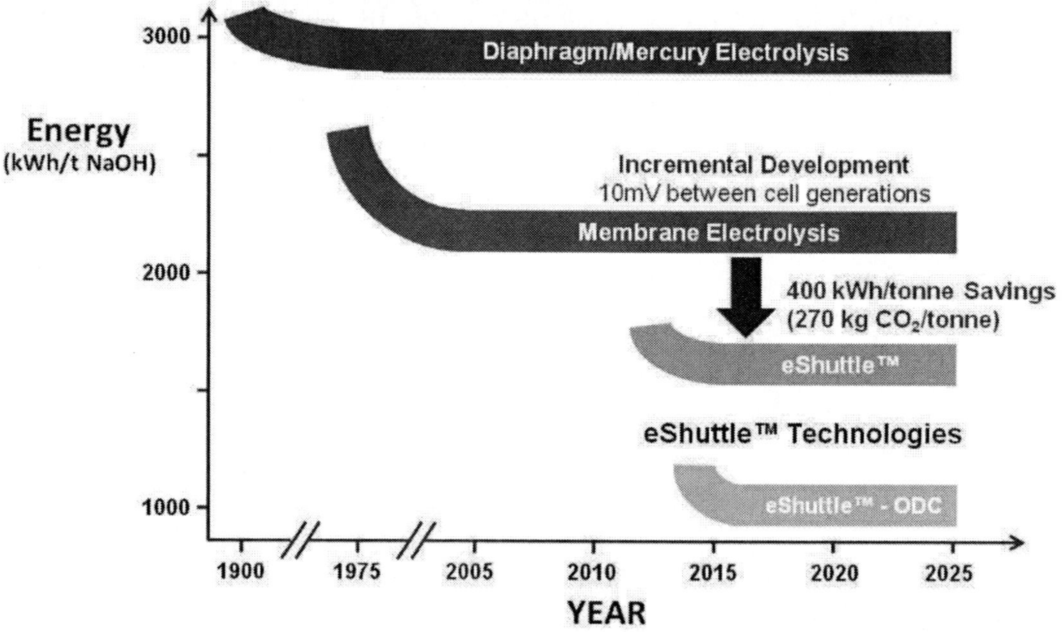

Figure 4: eShuttle™ is a significant advance in energy demand vs. traditional electrolysis routes[5].

eShuttle™ Process Thermodynamics:

The thermodynamics of the eShuttle™ process provide interesting insight into the source of the energy advantage over chlor-alkali/direct chlorination. Because both processes produce the same outputs from the same inputs, the minimum energy requirements are the same as defined by the Gibbs Free energy change for the overall reaction (see above):

$$2NaCl + 2H_2O + C_2H_4 \rightarrow C_2H_4Cl_2 + 2NaOH + H_2 \qquad \Delta G_{298K} = +274.5 \text{ kJ/mol}$$

[4] The reaction shown is for illustrative purposes. In fact, copper speciation is much more complex and the equilibrium speciation is dependent on a number of factors beyond the scope of this paper.

[5] The eShuttle™ ODC technology is the replacement of the traditional chlor-alkali cathode with an oxygen depolarized cathode (ODC). It is beyond the scope of this paper.

Because the beginning and ending states are the same, the energy savings must come from an improvement in efficiency. As will be discussed below, this efficiency improvement is fundamentally different than efficiencies achieved for the conventional chlor-alkali process, which have been small, engineering-based increments over a long period of time.

When $\Delta G > 0$, the Gibbs Free Energy change defines the minimum external work, electrical work in this case, that must be provided to a steady-state, continuous process. When the work is in the form of electricity, the above ΔG is better considered in electron-volts. If the reaction were a single-electron process, the minimum voltage required would be +2.84 V. Because the reaction is a two electron process, the minimum voltage for a cell supplying energy to the reaction is +1.42V (=2.84/2), a number that is quite close to the 1.5V theoretical voltage for the eShuttle™ process. Thus, the eShuttle™ process represents a very close approximation to the minimum required electrical energy for the combined processes. As with chlor-alkali, additional efficiency gains may still be realized through engineering improvements in membranes, cell design, and electrode performance, but these will be improvements on a superior, thermodynamically optimal baseline.

Another way of viewing the same information is shown in Figure 5 below. The y-axis is the Gibbs Free Energy of the system (based on the Gibbs Free Energies of formation for the components). The x-axis represents a nominal reaction coordinate. The top line follows the trajectory of the traditional chlor-alkali and direct chlorination processes, left side and right side respectively, and the lower line follows the trajectory for the eShuttle™ process.

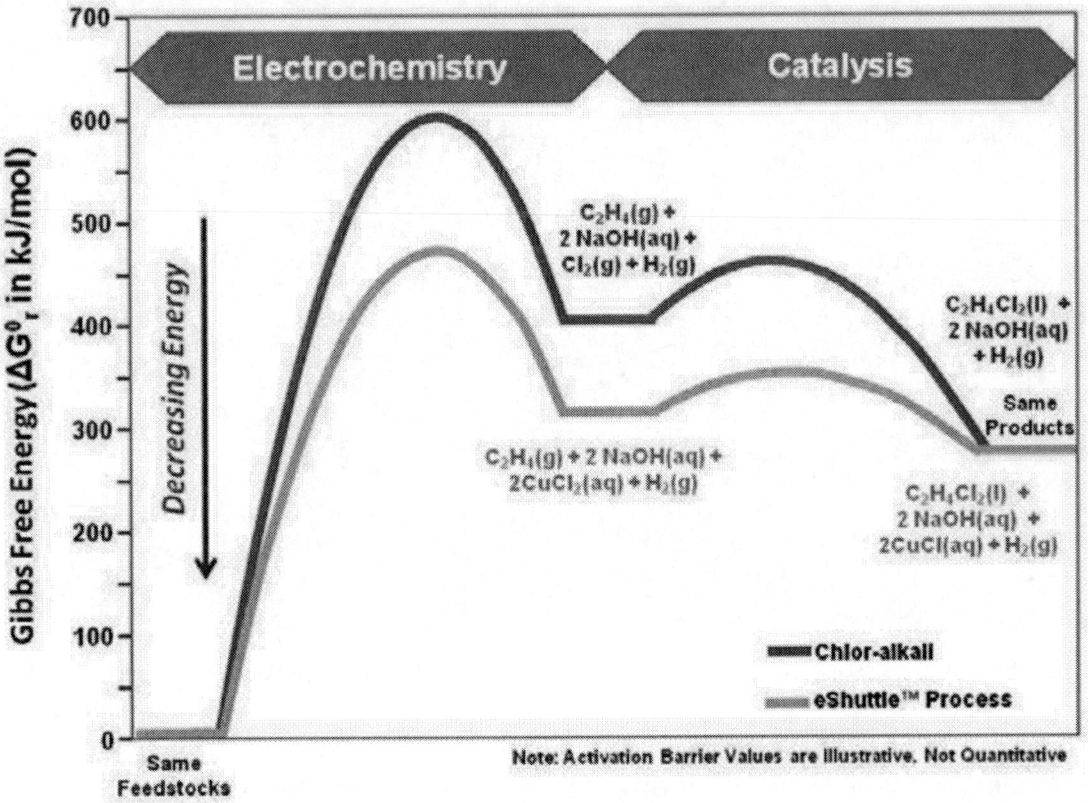

Figure 5: Thermodynamic comparison of chlor-alkali/direct chlorination and eShuttle™.

As noted previously, because both processes have the same inputs and outputs, the net change in going from left to right must be the same. As is clear in the above diagram, the voltage usage for the conventional chlor-alkali process leads to an unnecessarily high energy intermediate (i.e. chlorine). This extra energy can only be recovered as heat via downstream steam generation. Put another way, the steam recovered in direct chlorination is a direct result of the unnecessary voltage required in the chlor-alkali process. Because the intermediate state in the eShuttle™ process is much closer to the theoretical minimum electrical input, it represents the more efficient path.

Scale-up:

Over the past 36 months, the eShuttle™ process has been successfully implemented and tested at Chemetry's pilot scale facility in Moss Landing, CA. The pilot unit (shown below in Figure 6) is a fully automated plant designed with significant flexibility to test various process and equipment configurations. The electrochemical cell utilized is a full commercial height cell which allows for representative testing membranes and electrodes under commercial hydraulics. The packed bed process reactor is about 20 ft and can be operated in both trickle bed and bubble column flow regimes.

Figure 6: Chemetry's Moss Landing Pilot Plant. For scale, note the operators in the lower left corner.

During the operation, we have successfully demonstrated all phases of the integrated process including control systems, thermodynamic and transport data, component lifetimes, materials of construction, product quality, and environmental and safety systems.

Process Economics

The primary advantage for the eShuttle™ process is in the electricity savings. However, a full evaluation of the process economics must also account for differences in steam production and ethylene consumption between the combined chlor-alkali/direct chlorination processes and the eShuttle™ process. The economics should also account for regional price differences in these variables. As one example, conversion of an existing diaphragm-based chlor-alkali facility with direct chlorination to the eShuttle™ process would result in a net OpEx savings between $37/MT and $95/MT depending on regional costs as shown below in Table 1.

Table 1: Costs for electricity, steam and ethylene on a regional basis

Assumptions	Region/Market		
	US	Europe	SE Asia
Energy ($/MWh)	50	75	120
Steam ($/ton)	12	32	19
Ethylene ($/ton)	600	1100	1100
OpEx Savings $/MT EDC	**30**	**77**	**73**
OpEx Savings $/MT NaOH	**37**	**95**	**91**

As expected OpEx savings are greatest in regions where the power prices are highest. Capital equipment (CapEx) expenditures for the eShuttle™ process are comparable for chlor-alkali and direct chlorination in grassroots installations. As a result, the eShuttle™ process has the highest return on investment and is the most economical route to EDC.

Retrofit cases, which are beyond the scope of this paper, are better described on a case-by-case basis.

Conclusion

Replacing the standard chlor-alkali anode reaction for chlorine generation with a copper oxidation step reduces the cell operating voltage to near the theoretical minimum, which in turn leads to a dramatic reduction in overall power costs. As a result, Chemetry's eShuttle™ process provides the highest return on investment for EDC production when compared with conventional chlorine generation/direct chlorination technologies. In addition, the eShuttle™ process provides the added benefits of improved safety and lower carbon footprint. Over the past 36 months, the process has been demonstrated at the pilot plant scale at Chemetry's facility in Moss Landing, CA. Chemetry and TechnipFMC are actively pursuing demonstration plant opportunities as the next step in process commercialization.

479893 Application of LNG Technology Solutions for Optimisation of Liquefaction Processes

Abstract:

The application of proven, specialised equipment, efficient natural gas liquefaction process technology and modular construction is enabling the export of US natural gas by challenging the paradigm that a large baseload facility provides the best economy of scale. The concept is widely known as mid-scale LNG but the story doesn't have to end there. The principal reason for cryogenic liquefaction is storage and transportation and the same criteria that make the model work for LNG can be applied to other gases increasingly in demand; specifically ethylene, ethane, liquid air for energy storage, nitrogen, hydrogen and helium.

Key to cryogenic processing is the brazed aluminum heat exchanger (BAHX). A BAHX is a highly efficient, custom designed, compact heat exchange device that provides a heat transfer area significantly greater than conventional heat exchanger technology and also has an inherent multi-stream capability where 8 or more process streams in a single composite is common. Minimizing power consumption is a key design criterion for liquefaction processes, creating an inherent requirement for highly efficient heat exchange. Process efficiency is gained by minimizing temperature approach between the hot and cold streams. As the temperature approach decreases, the required surface area increases. With that in mind, the heat exchanger of choice must offer high surface area economically.

To ensure the robustness of the BAHX state-of-the-art software is coupled with in-house engineering expertise and experience to perform advanced transient thermal analyses. Exchangers are studied three-dimensionally from available operating data or dynamic simulation process software to determine the internal fluid and metal heat exchanger temperatures and the associated thermal stresses imparted on the exchanger. These analyses can be carried forward to predict potential fatigue damage and life expectancy of the equipment. Dynamic simulation is used to simulate operation of the BAHX and plant upset conditions to ensure process technologies are designed with proper controls to mitigate situations that introduce thermal stresses.

Typically refrigeration cycle configurations are focused on two; nitrogen cycle and mixed refrigerant cycle. Each cycle has its advantages and disadvantages when evaluated for use in a specific application. Usually designers study the trade-offs between power consumption and equipment utilization. This analysis allows them to optimize the liquefier for the total plant cost (capital plus operating costs). Comparison of operating costs should focus on the energy consumption of the refrigeration system, as the balance of plant should be essentially the same for all processes.

Nitrogen cycles are simple to operate and eliminate the need for the use and storage of hydrocarbons as refrigerants. They are ideally suited to remote areas that do not have easy access to hydrocarbons for refrigerants. However, simplification comes at the expense of higher specific energy consumption.

Mixed Refrigerant (MR) processes are based on mixtures of light hydrocarbons and nitrogen and process optimization is achieved by varying the mixture of refrigerant components and operating pressures in a manner that essentially allows the heating curve of the refrigerant to mirror the cooling curve of the gas to be liquefied in a tight temperature approach. The brazed aluminum exchanger accommodates the surface area requirements resulting from the tight approach temperatures and log mean temperature differences (LMTD). Its inherent multi-stream capability also allows the process engineer to add or remove refrigeration along the process thermal gradient, wherever needed, to optimize process efficiency.

Because all the capital equipment and technology required is already proven and deployed globally in LNG service, it represents a no risk solution in their application to other liquefaction processes.

A Process Synthesis Approach to Natural Gas Liquefaction

Baraka Celestin Sempuga, Diane Hildebrandt
Material and Process Synthesis (MaPS) University of South Africa (UNISA), PO Box 392, Unisa 0003, South Africa

Introduction

Natural gas is one of the most important commodities in the global market. Oil is the leading source of energy; however coal and natural gas contribute almost 50% of the energy source (S. A. Al-Sobhi et al., 2009) . Natural gas is becoming more attractive due to its environmental advantages compared to other fossil fuels such as coal and oil. CO_2 is produced from coal and oil at an approximate rate of 1.4 to 1.75 higher than that produced from natural gas. Furthermore NO_x produced from burning natural gas are approximately 20% less than those produced from burning coal and oil (Kidnay and Parrish, 2006). Natural gas is often found in remote locations and sometimes requires long distance transportation to utilization sites. It is either transported as a gas by pipeline or as Liquefied Natural Gas (LNG) using insulated tankers. The mode of transportation used is much dependent on the distance over which natural gas must be transported as well as on the cost of transportation.

LNG is produced by condensation of natural gas. Methane is the principle element of natural gas but may also contain ethane, propane and heavier hydrocarbons. Small quantities of nitrogen, oxygen, carbon dioxide, sulfur compounds and water may also be found in natural gas. The typical natural gas composition is approximately 82% methane. LNG is produced by a sequence of processing steps that transform natural gas from gas phase to liquid phase while getting rid of the undesirable substances. The final composition for LNG is usually determined by the market specifications(S. Al-Sobhi et al., 2009).

The LNG Process

The cooling and liquefaction of natural gas is generally done in a sequence of refrigeration cycles. The number of cycles is a key factor in the success of liquefaction process. To make the cold temperatures required for the LNG, work must be supplied to the refrigeration cycles and heat must be rejected from the cycles. The process is work intensive, and thus energy efficiency is important as feed gas is consumed in order to carry out the liquefaction process (Remeljej and Hoadley, 2006). Most of the existing LNG technologies use closed cycles, where external source of refrigeration is used to cool the natural gas in a number of steps until it liquefies. The external source of refrigeration uses a refrigerant (mixed or pure component) to create the cooling effect by compressing and expanding the refrigerant in a cycle. Among the known existing LNG technologies include APCI Propane pre-cooled mixed refrigerant process, Phillips optimized cascade process, Black & Veatch PRICO process, Statoil/Linge mixed fluid cascade process and Axens Liquefin.

The basic principle behind all the above mentioned LNG processes consists of matching as closely as possible the cooling and the heating curves of natural gas and the refrigerant, so as to reduce thermodynamic inefficiency. Figure 1 shows the cooling curve for natural gas and the heating curves for a refrigeration system using a mixed components refrigerant and a cascade refrigeration system using three different pure component refrigerants. We can see from Figure 1 that the mixed refrigerant system comes closest to the natural gas cooling curve and therefore results in reduced thermodynamic inefficiencies because it minimizes the temperature different between the two fluids compared to pure component refrigerants (Kidnay and Parrish, 2006).

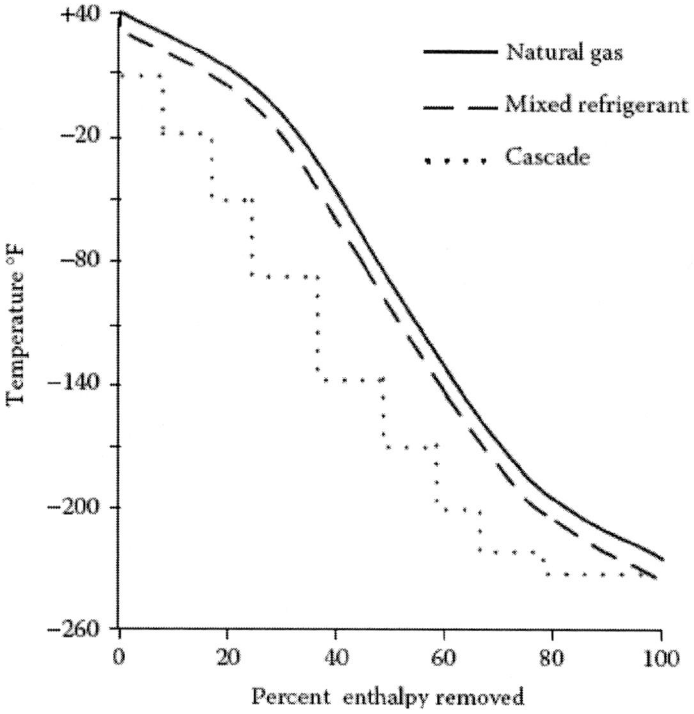

Figure 1: Cooling curve for natural gas and the corresponding heating curves for the mixed refrigerant and cascade cycle with pure component refrigerants (Kidnay and Parrish).

This manuscript presents an alternative way of analyzing LNG processes. The approach used is based on fundamental thermodynamic concepts and a graphical representation which can provide insight and information regarding the LNG process. The approach helps determining the targets for the process and can provide an idea of how these targets can be achieved. This approach is not only applicable to LNG processes but to any process that requires refrigeration.

LNG Process in the gh-diagram

It has been shown that Region 2 (that is 2A and 2B) of the gh-diagram (Figure 2) represents processes that run below ambient temperature and reject heat and require a refrigeration system. LNG is one of these processes as it requires refrigeration systems in order to liquefy natural gas. The ΔH axis of region 2 on the gh-diagram gives the amount of heat to be absorbed, the ΔG axis gives the minimum amount of work needed to absorb the heat and the temperature scale gives the reversible refrigeration temperature. This can be shown in Figure 2; a process located at point m (given by ΔH and ΔG across the process) would require that 14 kJ/mol of work be supplied, and this can be achieved by removing 10 kJ/mol of heat at 123 K; if this is done using a reversible refrigeration system then the net amount of work that would be consumed is 14 kJ/mol. The amount of work will increase as the temperature at which heat is removed decreases. For example at point n 10 kJ/mol of heat is removed at 80 K, therefore the minimum work required is 27 kJ/mol given by the intersection of line o-n and line m-n; we can see that if the same amount of heat needs to be removed at 0 K then an infinitely large amount of work will be required, this is shown by line o-q which can intercept line m-n only at an infinitely large value of ΔG.

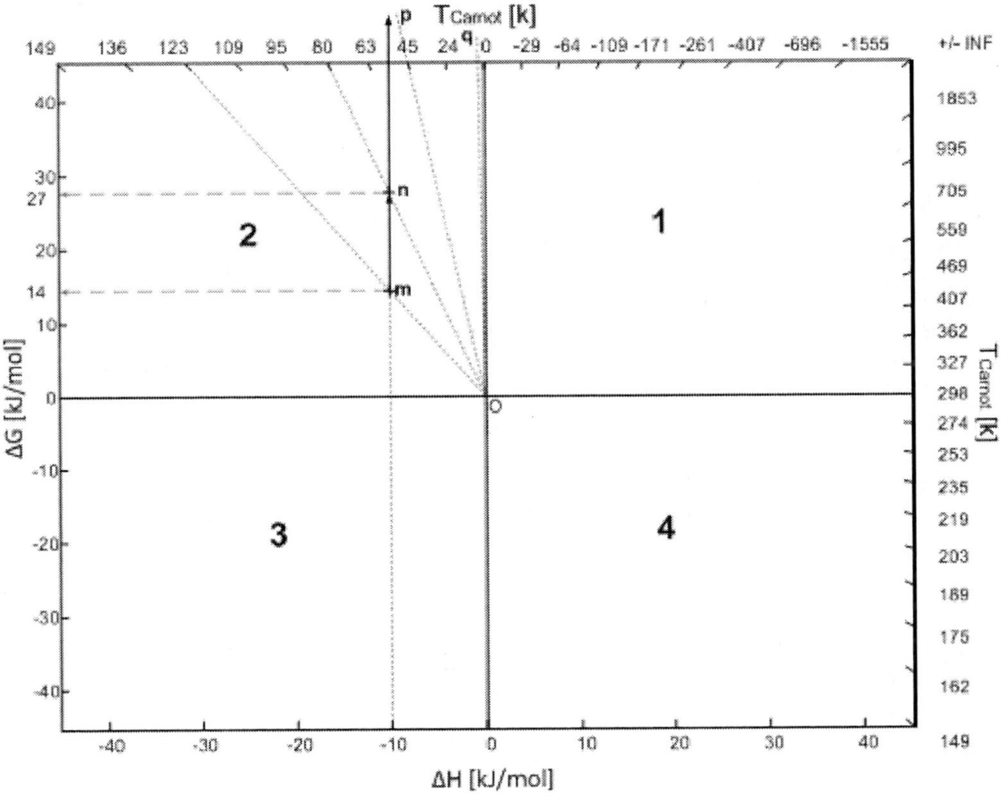

Figure 2: Example of a refrigeration process represented in region 2. Points m and n shows the amount of work needed (ΔG axis) to remove heat (ΔH axis) at the Carnot temperature (Temperature scale)

So far the gh-diagram has been used to represent processes where the input and output streams are at ambient conditions of temperature and pressure, T_O and P_O respectively. In this case the amount of work required in order for the process to proceed is given by the change in Gibbs free energy at T_O and P_O (ΔG (T_O, P_O)) (Denbigh, 1956). However at any other inlet and outlet temperature and pressure (T1, P1) and (T2, P2), the work required to transform A to B is given by the change in exergy (ΔE) across the process (Remeljej and Hoadley, 2006).

$$\Delta E = \Delta H\left(1 - \frac{T_O}{T}\right) \tag{1}$$

If the input and the output of the process are given, then ΔE and ΔH are known and therefore there is a unique value of T at which Equation (1) will hold. We will refer to this temperature as the Carnot temperature (Sempuga et al., 2010) it is the unique temperature at which the heat requirement must be supplied in order to satisfy the work requirement of the process. In a similar way as processes are represented in ΔH-ΔG diagram (gh-diagram), we can also represent processes in the ΔH-ΔE diagram (eh-diagram); here the input and output to the process are not restricted to ambient conditions. We can also identify the same thermodynamic regions in the eh-diagram as in the gh-diagram. We will now see that when we represent the LNG process in the eh-diagram, this can provide insight regarding the process and also provide a framework where alternatives can quickly be evaluated. Since natural gas is mainly composed of methane, in the following discussion we will use the thermodynamic properties of methane to represent natural gas. When we represent natural gas at various temperatures and pressures in the ΔH-ΔE domain this will form a surface which will be located in region 1 and region 2 of the eh-diagram. However, in this discussion we will only look at the part of the surface which is in region 2 as shown in Figure 3.

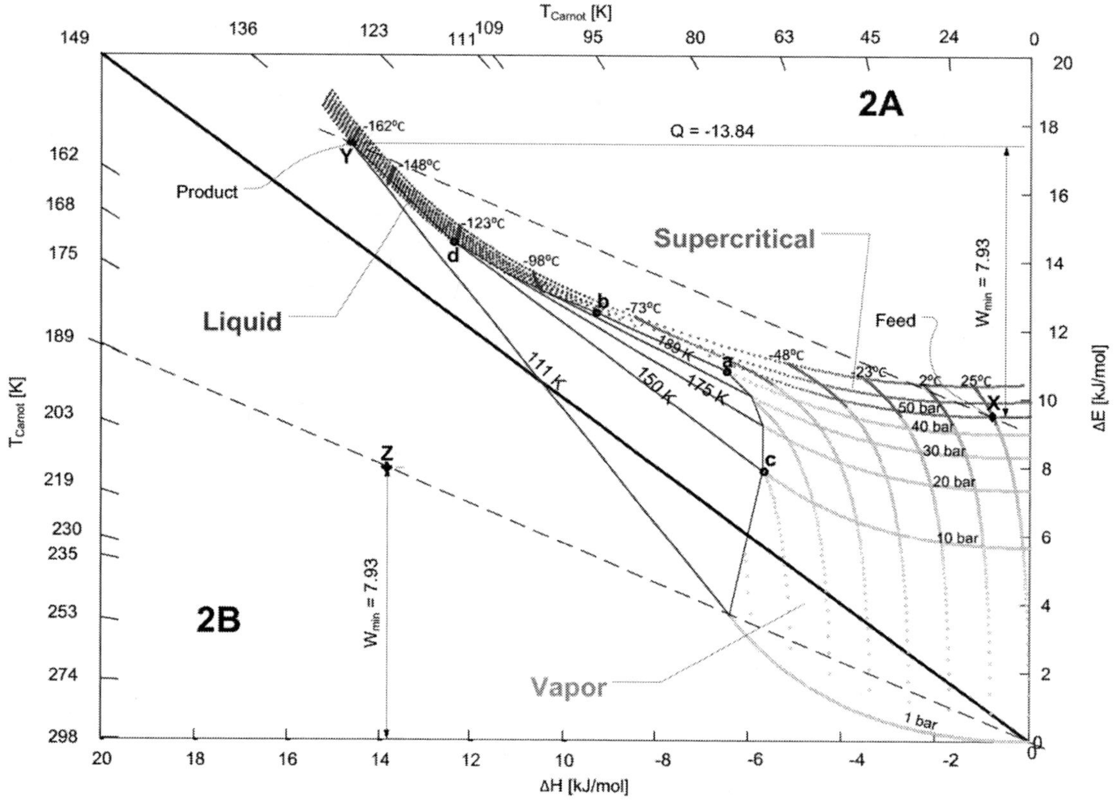

Figure 3: Natural gas at different temperatures and pressures represented on the eh-diagram. The basis is the natural gas enthalpy (H) and entropy (E) are zero at TO and PO. Colors are used to distinguish the different phases. Liquid, vapor, and supercritical

We will explore the usefulness of the eh-diagram (Figure 3) in the LNG process by looking at an example. Let us suppose that the feed to the process is natural gas (NG) at 25°C and 50 bar; this is located at point *X* on the eh-diagram in Figure 3. The product is liquefied natural gas (LNG) at -161°C and 1 bar; this is located at point *Y* on the eh-diagram. The minimum amount of work that the process requires to move from *X* to *Y* is given by the change along the ΔE axis and is W_{min} = 7.93 kJ/mol. This represents the work target for any route chosen to go from *X* to *Y*.

W_{min} can be supplied to the process by removing heat at the Carnot temperature. This can be seen on the eh-diagram in Figure 3; the Carnot temperature is obtained by subtracting *X* from *Y* to get point *Z*, and then the temperature is found at the intersection of line *O* – *Z* with the temperature scale. Note that line O – Z is parallel to line *X* – *Y*. Thus the Carnot temperature can be alternatively obtained by drawing a line parallel to *X* – *Y* from the origin to the temperature scale. This will give a temperature of 189 K (-84°C). This means that if we are able to remove the process heat (-13.83 kJ/mol) at the Carnot temperature (189 K), we would be able to reversibly liquefy NG. However we might need to manipulate the process stream to get it to the final state (111 K and 1 bar) by using heat exchangers and expanders and the process must be fully heat and work integrated internally such that only the process heat (-13.83 kJ/mol) at the Carnot temperature (189 K) is exchanged with an external refrigeration system as shown in Figure 4. On the other hand, in order for the refrigeration system to be reversible it must absorb the heat from the process at the process temperature and heat must be rejected from the refrigeration system at ambient temperature. In that case the refrigeration system will require an amount of work equivalent to W_{min}.

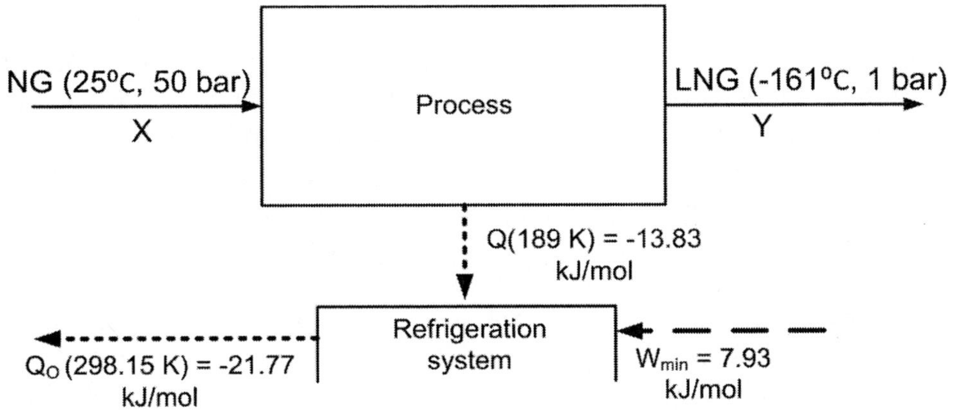

Figure 4: Schematic diagram Supplying work to a LNG process using a refrigeration system: Minimum work is supplied to the LNG process by removing heat at an appropriate temperature

Note that heat can be transferred at single temperatures only in the two-phase region of the natural gas envelop in Figure 3. Thus for the process in Figure 4, heat must be exchanged with the refrigeration system along line $a-b$ (at 189 K) on the eh-diagram (Figure 3) and the process route will be as follows: $X-a-b-Y$. However the process must be manipulated internally such that the amount of heat between $X-a$ and between $b-Y$ are rejected from the process only along line $a-b$; this might require a certain combination of recycles, heat exchangers, expanders and flashes as well as multiple points of heat transfers with the refrigeration system all happening at 189 K. Alternatively the process could be split into multiple processes using different refrigeration systems. For example the process could be split into three processes (Figure 4); the first process would go from X to c, the second from c to d and the third process, from d to Y. Each process with its own refrigeration system as shown in Figure 5.

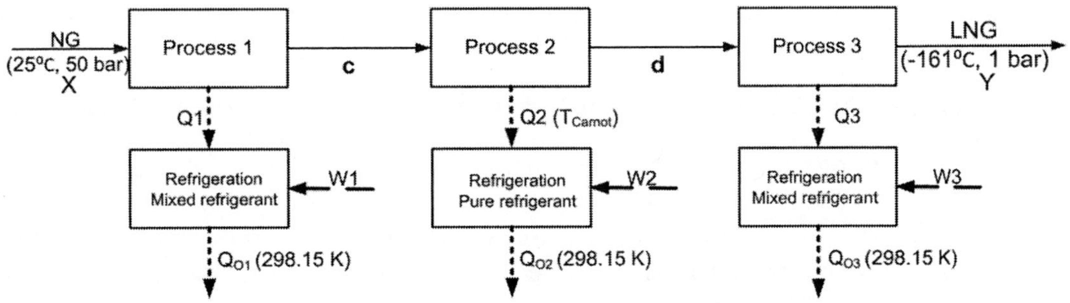

Figure 5: The LNG process split into three different processes, each with its own refrigeration system

Operations in Process 1 might involve expanding and cooling NG to move it from X to c. However cooling in this region must be done over a range of temperatures; this could be a source of irreversibility. In light of the diagram in Figure 1 it would be preferable to use a refrigeration system with a mixed refrigerant in order to reduce irreversibilities. Process 2 would operate in the two-phase region (*c* – *d*) where heat is removed at a single temperature. Thus in order for the process to be reversible, it would be preferable to use a refrigeration system with a pure refrigerant, since both the refrigerant and the process gas would change phase at a single temperature. Process 3 will also require cooling over a range of temperatures to move from *d* to *Y*. Thus in a similar way as in process 1, a mixed refrigerant system would be preferable in order to reduce irreversibilities.

Note that the temperature of any point in the two-phase region on the eh-diagram corresponds with its Carnot temperature; this can be seen in Figure 6: line *e* – *Y* is the vapor-liquid line of NG at 1 bar and 111 K, this line is parallel to line *O* – *p* indicating that each point of line *e* – *Y* has a Carnot temperature of 111 K. Similarly the vapor-liquid line at 10 bar and 150 K, line *c* – *d*, is parallel to line *O* – *q* indicating that each point on line *c* – *d* has a Carnot temperature of 150 K. This means that it is only in this region where work can be transferred reversibly to the process by removing heat at a single temperature and in one step. In other words, the two-phase region is the only region in the envelope where cooling can be done reversibly; thus a pure refrigerant system must be used in order to transfer heat at a single temperature. The other regions (liquid, vapor, and supercritical), will require cooling to be done over a range of temperatures, and therefore mixed refrigerants that approximate the cooling curves in these regions must be used in order to reduce irreversibilities.

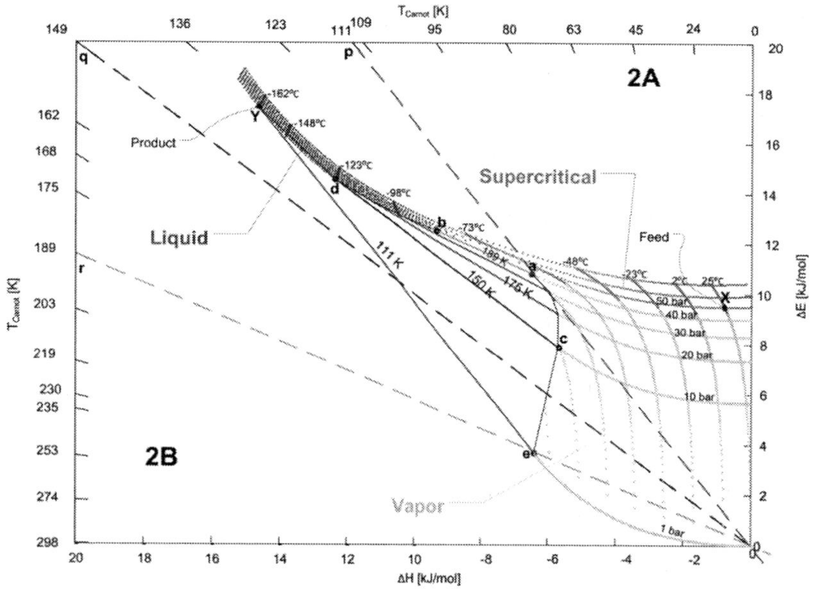

Figure 6: Natural gas on the eh-diagram: The temperatures in the two-phase region correspond to the Carnot temperatures

Furthermore minimum work requirement for refrigeration systems is achieved when the cycle operates in the two phase region at all times. Consequently the refrigerant must be partially evaporated before compression and partially liquefied before expansion. Irreversibilities occur when liquefaction and evaporation are not conducted at a single temperature.

Figure 7 (A) is a schematic representation of existing refrigeration cycles. The refrigerant in vapor phase is compressed to a pressure P_h and reaches a temperature T_{h1}; the gas is then liquefied by cooling it to a temperature T_{h2}. The liquefied refrigerant is then expanded, through a valve, to a pressure P_C where it evaporates partially. The mixed phase is then evaporated completely (usually at constant temperature) reaching a temperature T_C and pressure P_C before it is recompressed. The major source of inefficiency in these processes is in the liquefaction part since it occurs over a range of temperatures. We can show that the resulting temperature (T_{hot}) at which heat (Q_{out}) is removed during liquefaction is given by the log mean temperature between T_{h1} and T_{h2} (Equation (2)).

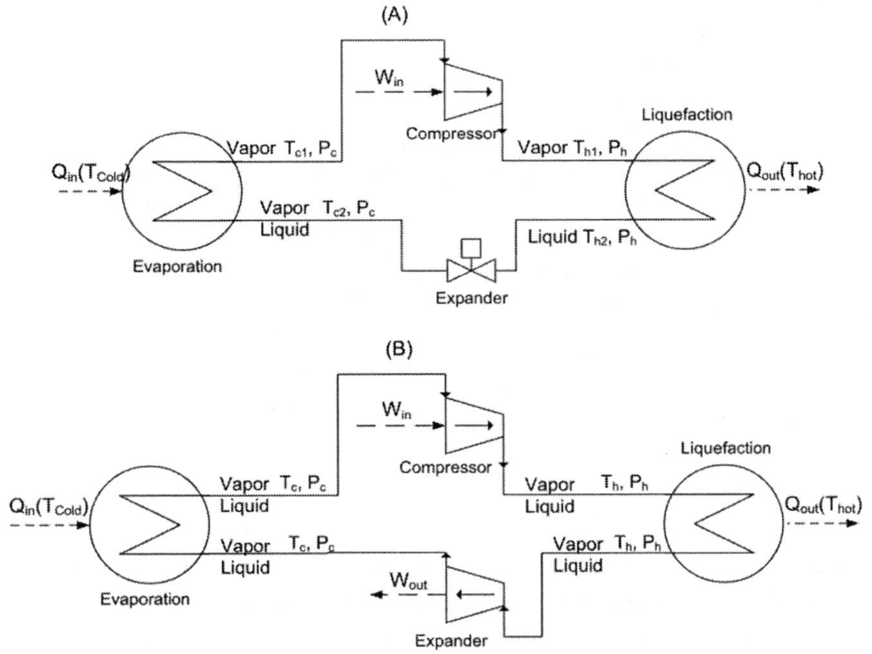

Figure 7: Schematic representation of refrigeration cycles. (A) In conventional refrigeration cycle the refrigerant is compressed in the vapor region consequently liquefaction and evaporation are conducted at multiple temperatures. (B) Compression and expansion are conducted in the two phase region consequently liquefaction and evaporation are conducted a single temperature and thus minimum work requirement is achieved

$$T_{hot} = \frac{(T_{h1} - T_{h2})}{\ln \frac{T_{h1}}{T_{h2}}} \tag{2}$$

In most cases irreversibility arises because T_{hot} does not match the Carnot temperature (T_{Carnot}) of the liquefaction unit. T_{Carnot} has been previously defined as the temperature at which heat must be transferred for the process to be reversible and can be obtained by rearranging Equation (1) as follows:

$$T_{Carnot} = \frac{T_O}{1 - \frac{\Delta E}{\Delta H}} \tag{3}$$

Where T_O is the ambient temperature and ΔE and ΔH are the change in exergy and change in enthalpy across the unit respectively.

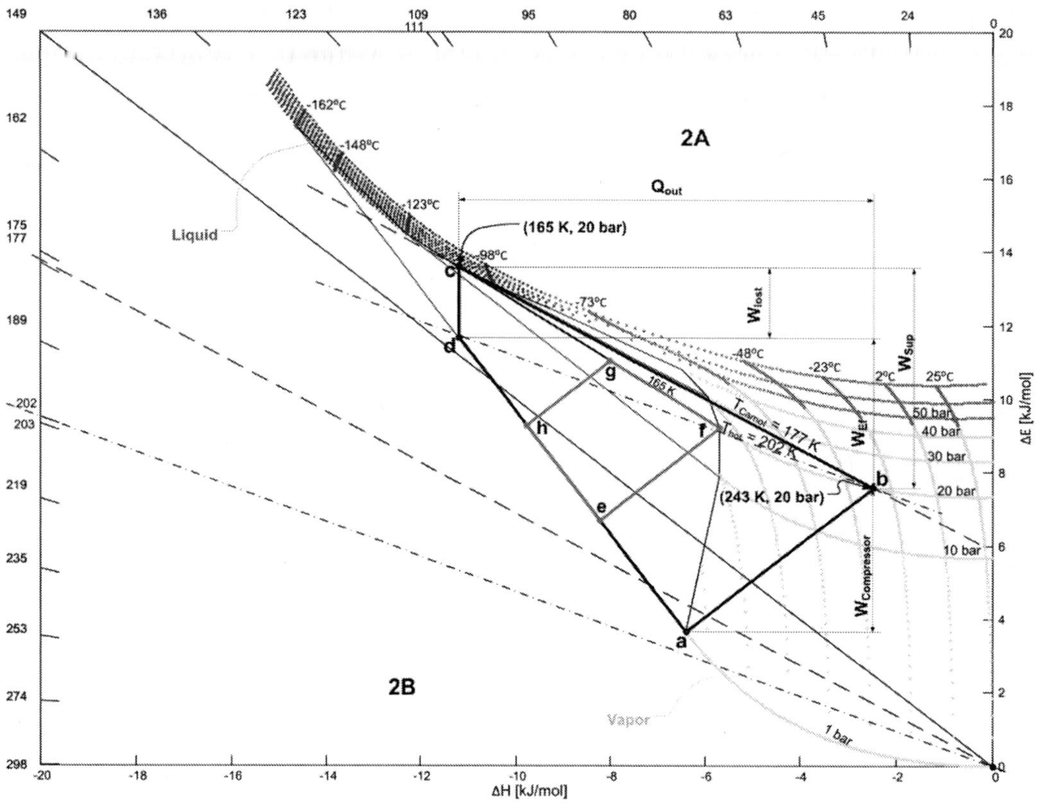

Figure 8: Refrigeration cycle on the eh-diagram. For path a-b-c-d, work is lost when cooling is done at multiple temperatures between b and c T_{Carnot} does not correspond to the actual temperature T_{QH} at which heat is removed, therefore more work (W_{Sup}) than necessary (W_{Ef}) is supplied to the process, and the process becomes irreversible. Every point in path e-f-g-h corresponds with the Carnot temperature; therefore no work is lost and the process is reversible.

As an illustration consider a refrigeration cycle represented on the eh-diagram in Figure 8 where the refrigerant is methane. The refrigeration path, a-b-c-d corresponds to the path on the diagram in Figure 7 (A). The gas is compressed isentropically from 1 bar (a) to 20 bar (b) reaching a temperature $T_{h1} = 243$ K. The gas is then cooled at constant pressure and liquefies at a temperature $T_{h2} = 165$ K (c). The Carnot temperature between b and c is $T_{Carnot} = 177$ K and the log mean temperature is $T_{hot} = 202$ K. We therefore see that the work supplied to move directly from b to c is $W_{Sup} = 6.01$ kJ/mol which correspond to heat ($Q_{out} = -8.75$ kJ/mol) being removed at 177 K. However in the actual process heat (Q_{out}) is being removed at 202 K and this corresponds to the effective work requirement of the process $W_{Ef} = 4.17$ kJ/mol. The work lost during liquefaction is $W_{lost} = 1.84$ kJ/mol. This work lost is the penalty for moving from b to c in one step. The work lost could be minimized if the gas is cooled progressively between b and f until it reaches the dew point, followed by liquefaction at constant temperature of 165 K.

On the other hand the refrigeration path e-f-g-h (Figure 8) is in the two phase region and corresponds to the path on the diagram in Figure 7(B). We can see that liquefaction and evaporation occur at constant temperatures 165 K and 111 K, and these temperatures correspond to their respective Carnot temperatures, therefore there is no work lost in the path. However to ensure that minimum work requirement is achieved one must use an expander that allows recovering the work of expansion which would otherwise be lost if a simple valve is used as in Figure 7(A) and thereby increasing the process irreversibility.

The major problem that could arise in implementing refrigeration in the two phase region is that most existing compressors and expanders operate efficiently only with vapor, and could result in major work losses when used in the two phase region. However screw compressors and expanders have shown to be able to handle two phase fluids at a relatively high efficiency (75%) (Smith et al., 2001). With further studies these equipment have the potential to be adapted to refrigeration systems. More studies (Hundseid et al., 2008) are also being conducted on centrifugal compressors and expanders for their possible efficient use in handling two phase fluids.

Conclusion

We have introduced a graphical approach to analyze LNG processes. The use of the eh-diagram can allow determining the optimum work target for the process and how this can be achieved as well as determine the optimum temperatures at which heat must be removed from the process. Furthermore the approach can be useful in scanning different routes from NG to LNG and help determine the appropriate refrigerant for each step of the process in a more systematic way. The eh-diagram can be used in the analysis and design of refrigeration cycles and can be used to match refrigerants in cascade refrigeration for optimum operation. We have also shown, by using the eh-diagram, that one of the possible ways for reaching minimum work requirement in refrigeration cycle is to operate in the two phase region at every point in the cycle. While most existing compressors and expanders cannot handle two phase fluids, numerous studies are being conducted in this regard and we see that the outcome of these studies could potential lead to much improved refrigeration systems.

References

Al-Sobhi, S.A., Alfadala, H.E., El-Halwagi, M.M., 2009. Simulation and Energy Integration of a Liquefied Natural Gas (LNG) Plant. In: Proceedings of the 1st Annual Gas Processing Symposium. pp. 131–135.

Al-Sobhi, S., Alfadala, H., El-Halwagi, M., 2009. Simulation and energy integration of a Liquefied Natural Gas (LNG) plant. In: Proceedings of the 1st Annual Gas Processing Symposium.

Denbigh, K.G., 1956. The second-law efficiency of chemical processes. Chem. Eng. Sci. 6, 1–9.

Hundseid, Ø., Bakken, L.E., Grüner, T.G., Brenne, L., Bjørge, T., 2008. WET GAS PERFORMANCE OF A SINGLE STAGE CENTRIFUGAL COMPRESSOR. In: Proceedings of ASME Turbo Expo 2008. pp. 1–10.

Kidnay, A.J., Parrish, W.R., 2006. Fundamentals of natural gas processing. CRC Press Tayllor Fr. 418.

Remeljej, C.W., Hoadley, a. F. a, 2006. An exergy analysis of small-scale liquefied natural gas (LNG) liquefaction processes. Energy 31, 1669–1683.

Sempuga, B.C., Hausberger, B., Patel, B., Hildebrandt, D., Glasser, D., 2010. Classification of chemical processes: A graphical approach to process synthesis to improve reactive process work efficiency. Ind. Eng. Chem. Res. 49.

Smith, I., Stosic, N., Kovacevic, A., 2001. Power recovery from low cost two-phase expanders. Trans. ….

485750 Smart Manufacturing Panel Discussion

Abstract:

Smart Manufacturing enables all information about the production/manufacturing process to be available when it is needed, where it is needed, and in the form it is needed across the entire manufacturing value-chain to power smart decisions. Learn about future of production and manufacturing from Smart Manufacturing experts. Hear firsthand experiences about setting strategies, supporting technologies, and lessons from implementation.

479973 The Ergun Equation: Right or Wrong?

Abstract:

When it comes to fluid flow through packed beds, the Ergun equation is the most popular model for relating fluid pressure drop to flowing fluid Reynolds number. The popularity of this equation results from it being

- presented in most engineering texts concerned with such flow;
- taught as "the gospel" in most top-tier engineering schools.

Thus, future engineers leave university with the Ergun equation well-represented in their intellectual toolkit.

Unfortunately, the Ergun equation rests upon shallow foundations laid in shifting sands. Basically, the Ergun equation is nothing more than the addition of the Blake-Kozeny and Burke-Plummer equations, both of which contain a plethora of questionable assumptions.

This presentation reports an experimental test of the Ergun equation and discusses the theoretical foundations of the Ergun equation.

This presentation also contains a "new look" at fluid flow through packed beds and demonstrates how to scale packed beds from laboratory-sized equipment to commercial-sized equipment.

480821 Product Development Challenges for Small or Start-up Companies

Abstract:

Product development challenges for small or start-up companies

Pratik C Pednekar, Research Chemical Engineer, Mid-Atlantic Technology Research and Innovation Center
pratik.pednekar@matricinnovates.com

John Dever, Executive Vice President, Mid-Atlantic Technology Research and Innovation Center
jack.dever@matricinnovates.com

Development is a key stage towards the commercialization of a product or a process. For a startup company or a department that is still in the process of looking for seed money, and is currently working with limited resources, this step becomes even more crucial. In order to engage potential investors, a small scale demonstration of the technology or a scale up of the product to send samples becomes a limiting step due to the investment required to perform this. Setting up a facility from the ground up may not be an option for many such companies due to cost or risk reasons. Other hurdles include the need to hire specialized vs flexible manpower, to have continuous access to expensive analytical tools and to be in touch with market experts who have worked with products that may put the company through a slow growth curve.

Another key aspect of this stage is the flexibility to handle growth of the project. For a company developing a candidate product, the goals for a project may change very rapidly from producing a defined amount of product to optimizing the product properties to being able to demonstrate continuous production or sustained quality. These changes can make some of the initially obtained equipment or workforce obsolete through the course of development and these are added costs towards a necessary path of progress for the company. Additionally after all this effort, it is quite possible that the product or technology may not be cost effective or desirable at this point in time. To know this information as early as possible through the development stage can make a huge difference to the survival of a start-up company.

Getting in touch with a facility that has most of the infrastructure already in place can be vital to help the project get off the ground quicker and cost effectively. This could save on costs for setting up a facility that is compliant for production, procuring the right equipment quickly and obtaining the manpower to build, operate and guide the development process. Through university or company collaboration or through access to the consultation entities, the development stage can be streamlined to the benefit of the start-up company. However, each of these options differ in several ways such as their access to resources, industrial contacts, intellectual property agreements, ability to adapt the manpower according to the growth of the project etc. Through working with start-up companies previously, MATRIC would like to present a discussion on some of the challenges that such companies face and how technology development companies with such facilities can help make the product development phase cost effective and efficient. The discussion is intended to demonstrate some of the cost saving means that can be employed through the development stage to reduce risk and at the same time provide access to a flexible workforce that can be adjusted based on the growth of the project.

479178 Scaling Packed Bed Columns and Reactors

Abstract:

In 1952, Sabri Ergun published an article reconciling all the data pertaining to flow through packed columns at that time. The equation presented in that paper has since become known as the Ergun equation. It has been used with varying degrees of success since its publication. The Ergun equation has various problems. First, it uses a Reynolds number based on particle diameter, not fluid conduit diameter. Thus, the Reynolds number used in the Ergun equation does not describe fluid flow. Second, the Ergun equation was developed for spheres. However, myriad odd-shaped packing exists; thus, the Ergun equation must be modified when using non-spherical packing. That modification is generally encapsulated in a dimensionless "sphericity factor" ... a cheat factor, in other words. Third, the Ergun equation provides no ready mechanism for upscaling or downscaling a process.

In this paper, we address these problems via Dimensional Analysis. We use Dimensional Analysis to formulate a new method for characterizing flow through packed columns. We also propose a new method for characterizing the packing material. Since our results obey the laws of similarity, our analysis can be used for upscaling and downscaling packed columns, which we will demonstrate in this presentation.

481063 Research Challenges in the Light Component Separation for the Methane Future Feedstock Projects

Abstract:

Methane as an alternative feedstock has been studied in the research division for many years. Most of the projects are diminishing as the challenges of reducing high process cost has been increased with the low ethane cost on the market in these years. Methane reformer is one of the successful technology that has been implemented for a few commercial routes in basic chemical industries. Other technologies, like OCM, MTM, MCTO, and combustion pyrolysis, are all under the research and development stages.

In this communication, the authors would like to share the overviews and to discuss the main challenges of these technologies. In particular, for the MCTO technology, the authors will introduce an innovative process design for the light hydrocarbon separation and an optimal integrated process to identify the major milestones required for the future commercial success.

INHA UNIVERSITY

Modelling and Optimization of Environment-Friendly Process at Oil Offshore Platform

2017.03.27

Graduate School of Chemistry and Chemical Engineering, Inha University

Yongheon Cho, Soojin Kwon, Junghoon Kim and Sungwon Hwang

Contents

1. Backgrounds
2. Modeling and Simulation
3. Optimization procedure
4. Optimization results
5. Summary

1. Backgrounds

1. Background

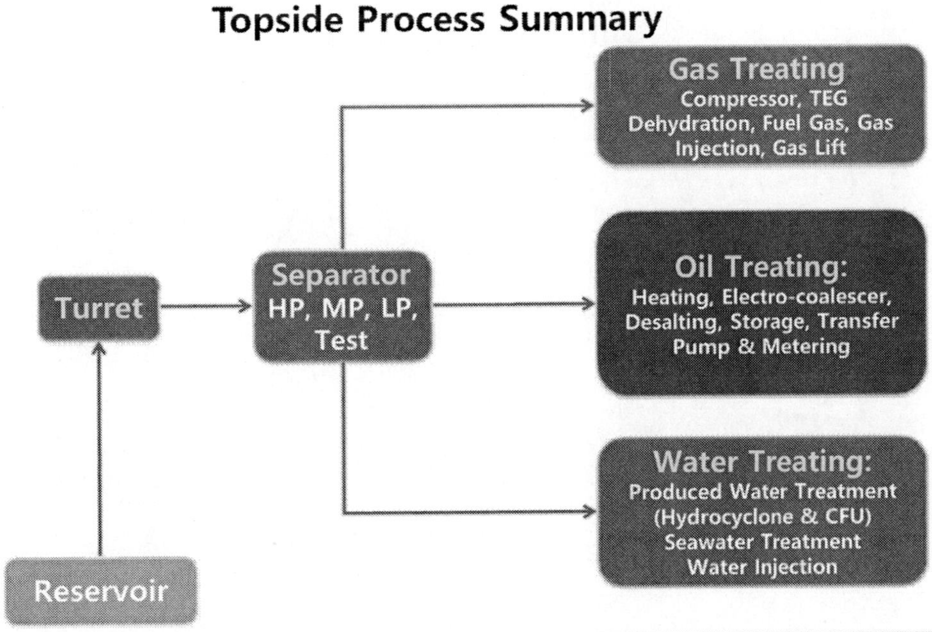

2. Modeling and Simulation

2. Modeling and Simulation

Simulation of Case 1

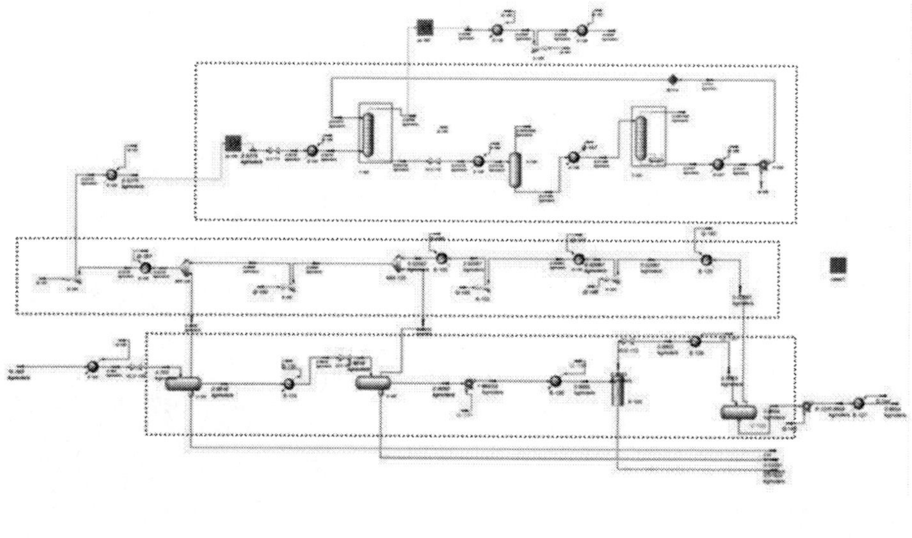

2. Modeling and Simulation

Case 1 : 3 separators without scrubber

- Use 3 separators
- Gas separated from separator is compressed without condensate recycling
- Variables : Pressures of V-102, V-103 and discharge pressure of K-101, K-102 and K-103

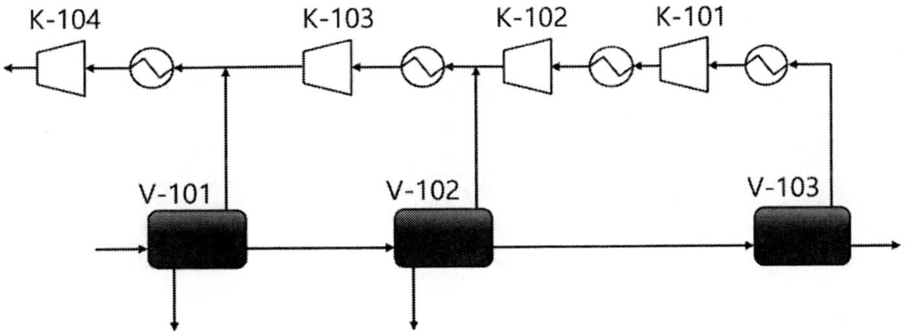

2. Modeling and Simulation

Simulation of Case 2

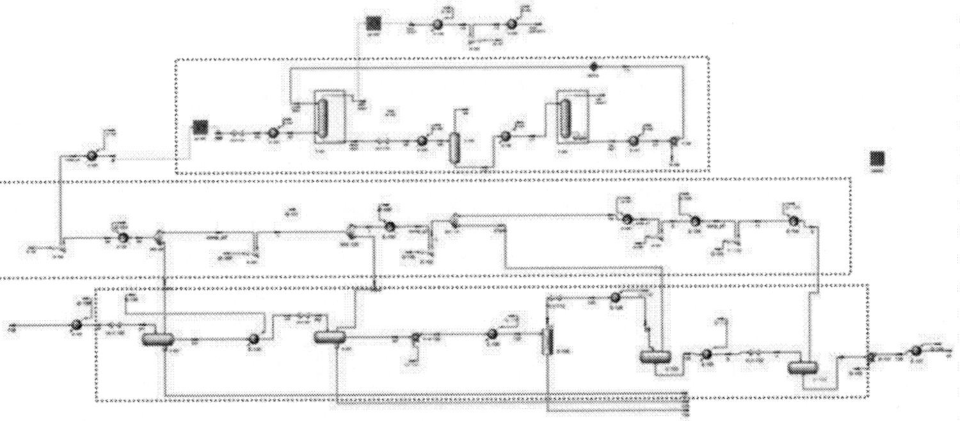

2. Modeling and Simulation

Case 2 : 4 separators without scrubber

- Use 4 separators
- Gas separated from separator is compressed without condensate recycling
- Variables : Pressures of V-102, V-103 , V-104 and discharge pressures of K-101, K-102, K-103 and K-104

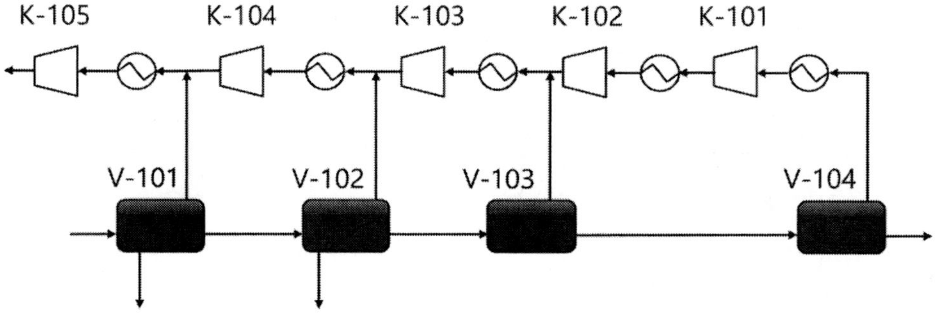

2. Modeling and Simulation

Simulation of Case 3

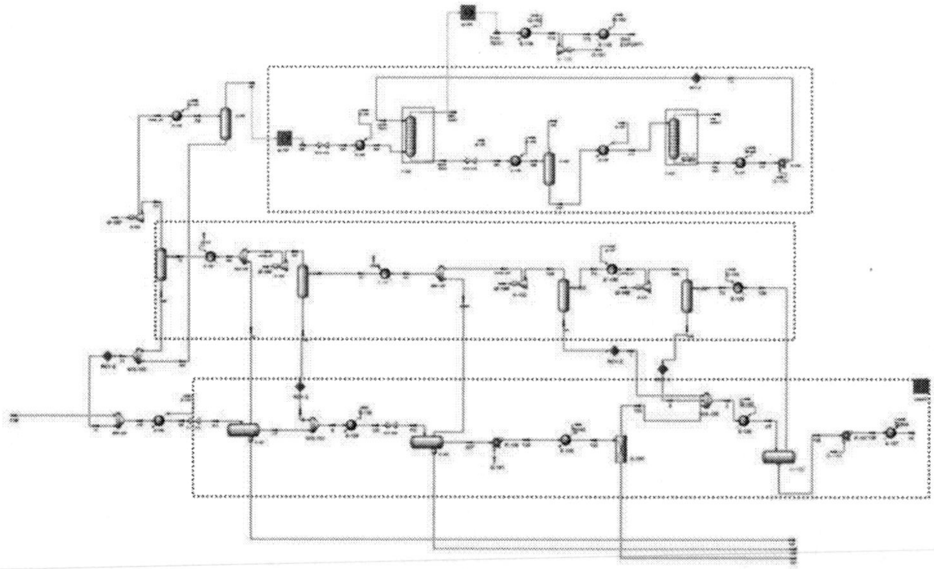

2. Modeling and Simulation

Case 3 : 3 separators with recycling

- Use condensate scrubbers
- Gas separated from separator is compressed with condensate recycling
- Operating variables : Pressures (V-102 and V-103) discharge pressures (K-101, K-102, K-103, K-104), Temperatures (V-201, V-202, V-203 and V-204)

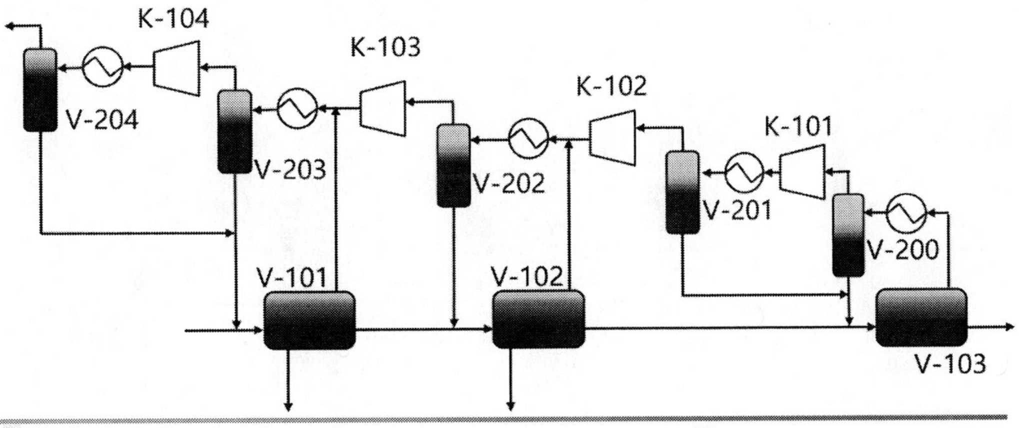

2. Modeling and Simulation

Simulation of Case 4

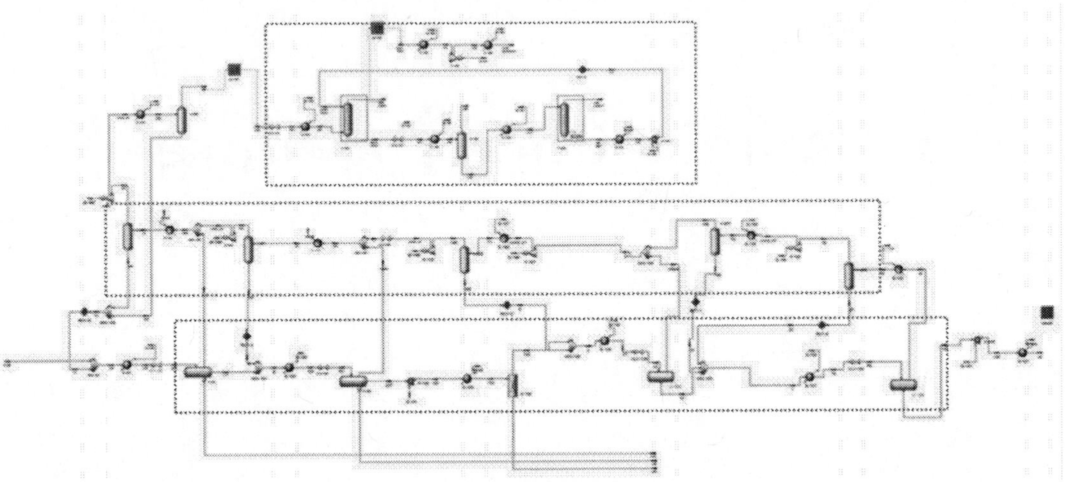

2. Modeling and SImulation

Case 4 : 4 separators with recycling

- Use condensate scrubbers
- Gas separated from separator is compressed with condensate recycling
- Variables : Pressures (V-102, V-103, V-104), discharge pressure (K-101, K-102, K-103 and K-104), Temperatures (V-201, V-202, V-203, V-204 and V-205)

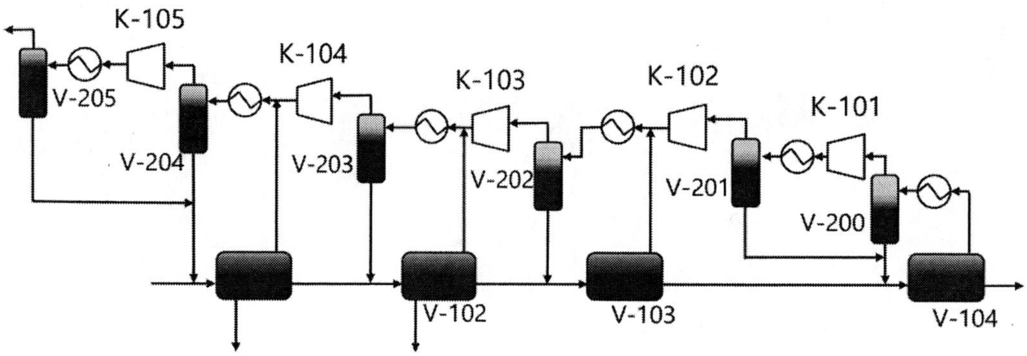

INHA UNIVERSITY 13 Clean Energy Process Integration Lab

INHA UNIVERSITY

3. Optimization procedure

3. Optimization procedure

1. Simulation Based Optimization Method

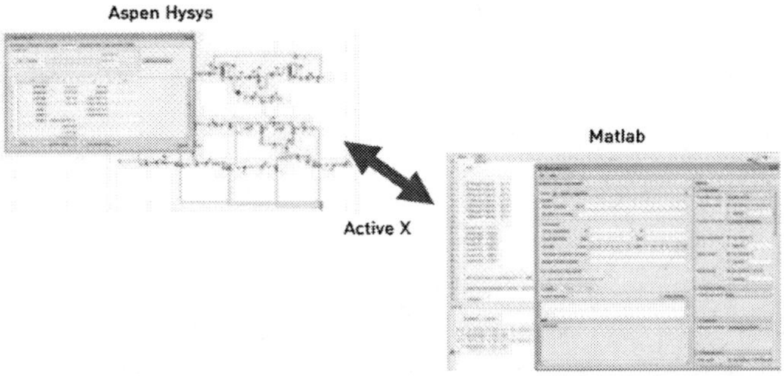

GA : Based on the mechanics of biological evolution

A simulated population of randomly selected individuals is generated and then allowed to evolve.

Express the problem in terms of a bit string
X = (1001010101011100)

Assign a fitness to each individual and Repeat until n children have been produced

3. Optimization procedure

2. Optimization Algorithm

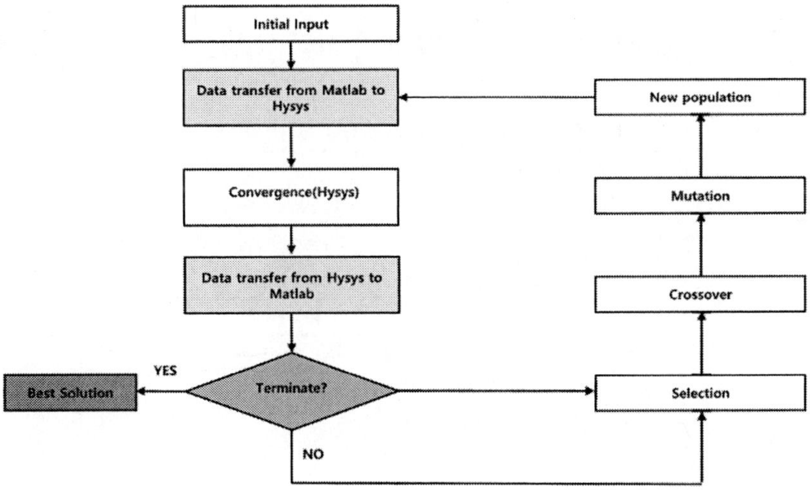

3. Optimization procedure

3. Operating Variables and Bounds

		case1			case2			case3			case4		
	notation	peak oil	peak gas	peak water	peak oil	peak gas	peak water	peak oil	peak gas	peak water	peak oil	peak gas	peak water
1st separator pressure	P1 (bar)	30	30	30	30	30	30	30	30	30	30	30	30
2nd separator pressure	P2 (bar)	12 - 30	12 - 30	12 - 30	12 - 30	12 - 30	12 - 30	12 - 30	12 - 30	12 - 30	12 - 30	12 - 30	12 - 30
3rd separator pressure	P3 (bar)	1 ~ 6	1 ~ 11	1 ~ 6	6 ~ 20	10 ~ 20	6 ~ 20	1 ~ 6	1 ~ 11	1 ~ 11	6 ~ 20	10 ~ 20	10 ~ 20
4th separator pressure	P4 (bar)	-	-	-	1 ~ 6	1 ~ 11	1 ~ 6	-	-	-	1 ~ 6	1 ~ 11	1 ~ 11
1st separator temperature	T1	29.8	59.2	29.9	29.8	59.2	29.9	29.8	59.2	29.8	29.8	59.2	29.9
2nd separator temperature	T2	60	60	60	60	60	60	60	60	60	60	60	60
3rd separator temperature	T3	72	72	72	72	72	72	72	72	73	72	72	72
4th separator temperature	T4	-	-	-	85	85	85	-	-	-	85	85	85
compressor P1	comP1(bar)	1 ~ 7	1 ~ 11	1 ~ 7	1 ~ 9	1 ~ 11	1 ~ 9	1 ~ 7	1 ~ 11	1 ~ 11	1 ~ 9	1 ~ 12	1 ~ 11
compressor P2	comP2(bar)	7 - 20	11 - 20	7 - 20	9 - 20	11 - 20	9 - 20	7 - 20	11 - 20	11 - 20	7 - 20	12 - 20	11 - 20
compressor P3	comP3(bar)	20 - 30	20 - 30	20 - 30	9 - 20	11 - 20	9 - 20	20 - 30	20 - 30	20 - 30	7 - 20	12 - 20	11 - 20
compressor P4	comP4(bar)	75	75	75	20 - 30	20 - 30	20 - 30	75	75	75	20 - 30	20 - 30	20 - 30
compressor P5	comP5(bar)	-	-	-	75	75	75	-	-	-	75	75	75
scrubber temperature 1	scr_T1 (℃)	-	-	-	-	-	-	38 ~ 60	38 ~ 60	38 ~ 60	38 ~ 60	38 ~ 60	38 ~ 60
scrubber temperature 2	scr_T2 (℃)	-	-	-	-	-	-	38 ~ 60	38 ~ 60	38 ~ 60	38 ~ 60	38 ~ 60	38 ~ 60
scrubber temperature 3	scr_T3 (℃)		-	-	-	-	-	38 ~ 60	38 ~ 60	38 ~ 60	38 ~ 60	38 ~ 60	38 ~ 60
scrubber temperature 4	scr_T4 (℃)	-	-	-	-	-	-	38 ~ 60	38 ~ 60	38 ~ 60	38 ~ 60	38 ~ 60	38 ~ 60
scrubber temperature 5	scr_T5 (℃)	-	-	-	-	-	-	-	-	-	38 ~ 60	38 ~ 60	38 ~ 60

3. Optimization procedure

4. Objective function : Maximize profit

$$\text{Max } f_{profit}(x) = Sales_{oil}(x) + Sales_{gas}(x) - Cost_{utility}(x)$$

Prices of Oil, Gas and utilities

item	price
gas product	$3.047/MMBtu
oil product	$43/bbl
steam(100pisg)	$7.42/1000lb
electricity	$0.1/kWh

* From U.S Energy Information Administration.
* Cooling water expense was ignored.

INHA UNIVERSITY

4. Optimization results

4. Optimization results

Maximum Profit and Optimum Operating Conditions

Peak oil result

	notation	case1	case2	case3	case4
2nd separator pressure	P2 (kPa)	1835.4	1864.9	1769.2	1868.8
3rd separator pressure	P3 (kPa)	515.4	695.4	513.3	671.2
4th separator pressure	P4 (kPa)	-	548.5	-	538.2
compressor P1	comP1(kPa)	532.4	681.5	560.8	567.1
compressor P2	comP2(kPa)	782.2	1130.5	1675.9	1343.3
compressor P3	comP3(kPa)	2252.3	1863.9	2999.8	1864.9
compressor P4	comP4(kPa)	-	2934.5	-	2988.0
scrubber temperature 1	T1 (℃)	-	-	38.3	41.9
scrubber temperature 2	T2 (℃)	-	-	38.3	38.0
scrubber temperature 3	T3 (℃)	-	-	38.3	38.4
scrubber temperature 4	T4 (℃)	-	-	58.1	38.2
scrubber temperature 5	T5 (℃)	-	-	-	56.5
steam cost	$/day	42422.0	62438.6	42476.3	62518.1
electric cost	$/day	52319.5	51513.6	49482.8	49371.2
oil product sales	$/day	27847786.8	27847987.6	27897460.2	27897297.5
gas product sales	$/day	532768.6	532762.2	533410.7	533666.7
overall profit	$/day	28285813.9	28266797.5	28338911.7	28319075.0
oil RVP	kPa	68.0	67.8	67.8	66.6

4. Optimization results

Maximum Profit and Optimum Operating Conditions

Peak gas result

	notation	case1	case2	case3	case4
2nd separator pressure	P2 (kPa)	2444.0	2410.1	1977.0	2475.6
3rd separator pressure	P3 (kPa)	950.9	1367.6	937.5	1324.1
4th separator pressure	P4 (kPa)	-	990.4	-	987.2
compressor P1	comP1(kPa)	984.9	990.9	960.4	1198.4
compressor P2	comP2(kPa)	1995.5	1424.5	1635.2	1474.0
compressor P3	comP3(kPa)	2967.2	1955.9	2999.7	1761.7
compressor P4	comP4(kPa)	-	2995.0	-	2976.7
scrubber temperature 1	T1 (℃)	-	-	38.6	42.9
scrubber temperature 2	T2 (℃)	-	-	38.0	41.7
scrubber temperature 3	T3 (℃)	-	-	38.4	41.3
scrubber temperature 4	T4 (℃)	-	-	60.0	38.4
scrubber temperature 5	T5 (℃)	-	-	-	58.6
steam cost	$/day	10783.2	35443.3	11216.3	35947.2
electric cost	$/day	49572.8	47549.1	45843.5	45987.5
oil product sales	$/day	33453293.1	33452471.8	33575109.0	33575180.4
gas product sales	$/day	510671.0	510802.8	512648.0	512659.3
overall profit	$/day	33903608.0	33880282.0	34030697.0	34005905.0
oil RVP	kPa	68.0	67.6	68.0	68.0

4. Optimization results

Maximum Profit and Optimum Operating Conditions

Peak water result

	notation	case1	case2	case3	case4
2nd separator pressure	P2 (kPa)	1362.6	2233.4	2425.1	2492.0
3rd separator pressure	P3 (kPa)	464.3	909.1	819.0	1061.3
4th separator pressure	P4 (kPa)	-	500.5	-	872.7
compressor P1	comP1(kPa)	569.0	701.3	1074.7	1045.4
compressor P2	comP2(kPa)	1329.6	1997.4	1799.7	1377.5
compressor P3	comP3(kPa)	2977.9	1252.9	2999.4	1703.7
compressor P4	comP4(kPa)	-	2999.8	-	2995.3
scrubber temperature 1	T1 (℃)	-	-	40.1	54.1
scrubber temperature 2	T2 (℃)	-	-	39.3	50.6
scrubber temperature 3	T3 (℃)	-	-	38.2	44.0
scrubber temperature 4	T4 (℃)	-	-	56.0	38.6
scrubber temperature 5	T5 (℃)	-	-	-	55.0
steam cost	$/day	7088.9	9803.2	7197.0	9934.8
electric cost	$/day	33923.1	33811.2	32181.1	32194.6
oil product sales	$/day	3699292.4	3699305.4	3748972.6	3747785.0
gas product sales	$/day	369463.5	369455.2	369233.2	369161.0
overall profit	$/day	4027744.0	4025146.0	4078828.0	4074817.0
oil RVP	kPa	67.9	67.9	66.5	67.1

4. Optimization results

Summary of Overall Profit - peak oil

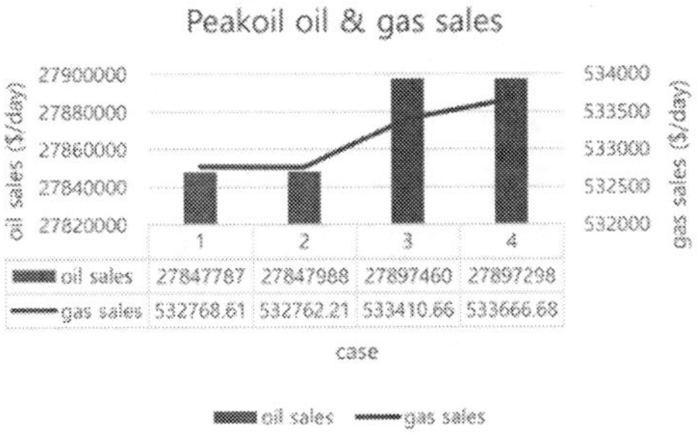

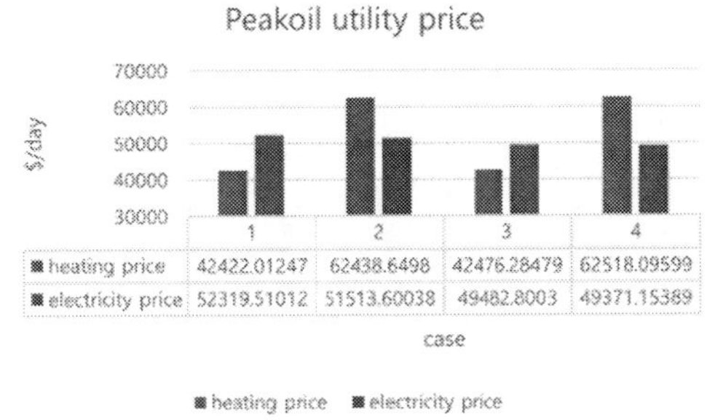

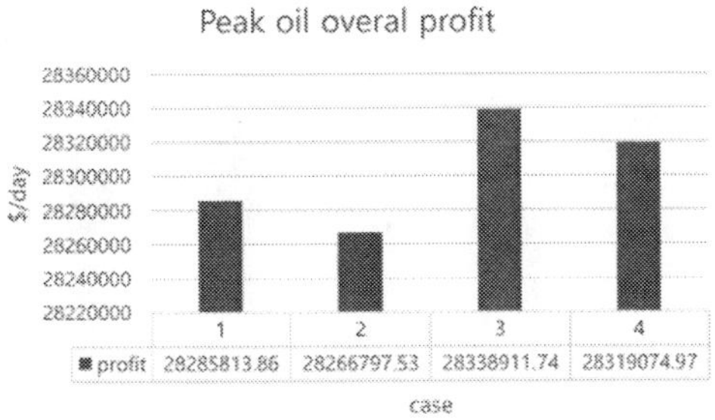

INHA UNIVERSITY 23 Clean Energy Process Integration Lab

4. Optimization results

Summary of Overall Profit - peak gas

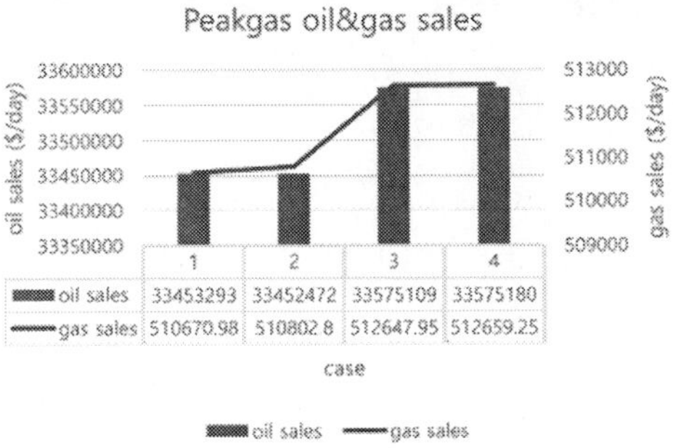

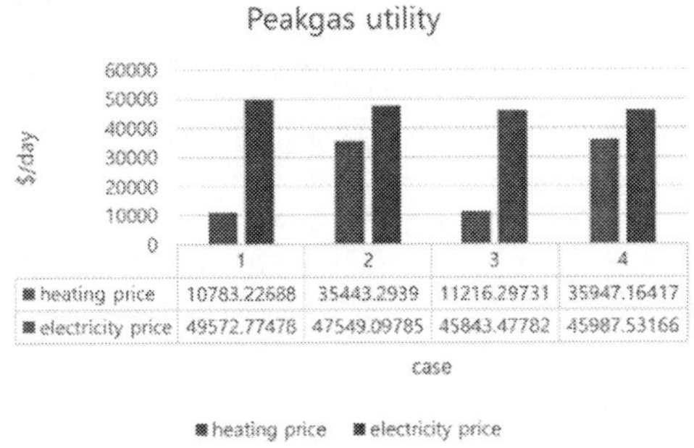

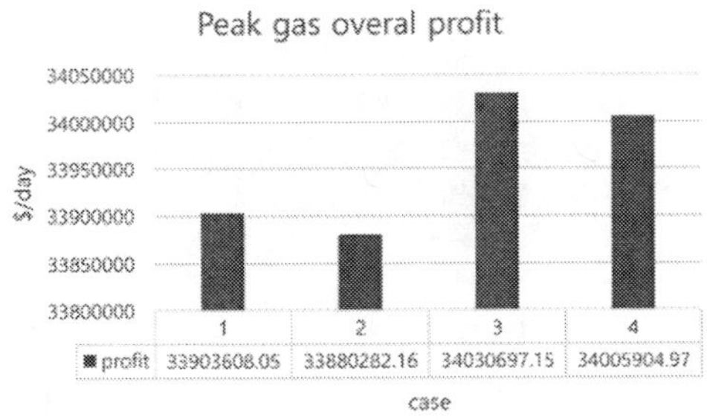

INHA UNIVERSITY 24 Clean Energy Process Integration Lab

4. Optimization results

Summary of Overall Profit - peak oil

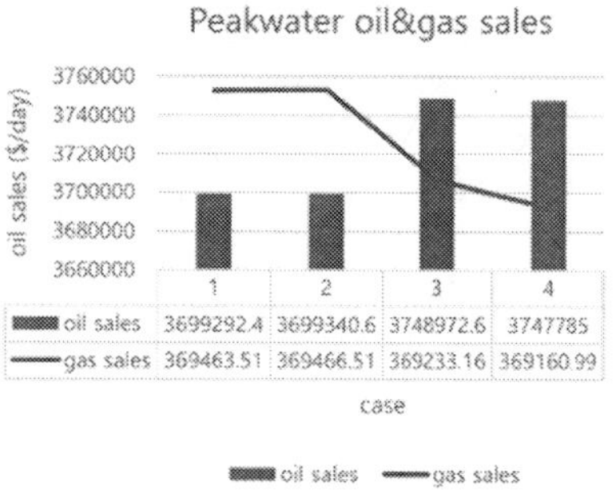

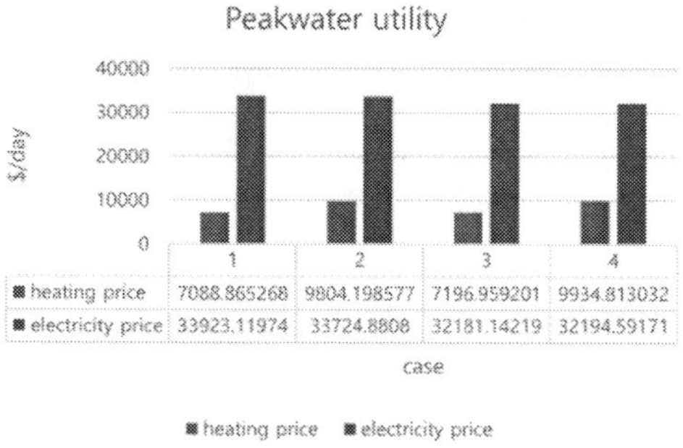

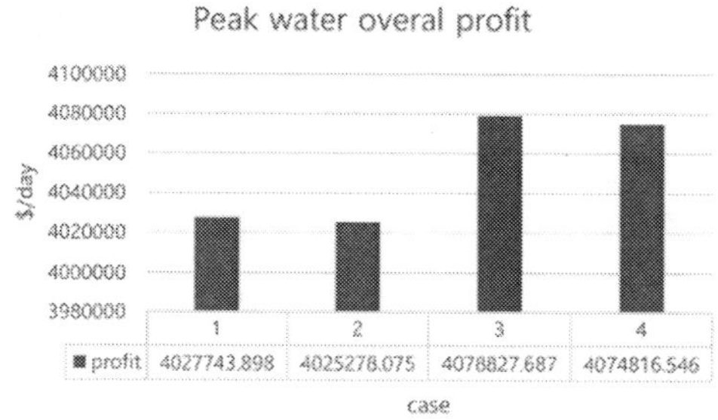

4. Optimization results

Summary of Capital cost

	peakoil	peakgas	peakwater
case1	34635100	42132100	25384600
case2	37942900	45539900	25775800
case3	35000000	36362000	25781100
case4	39394500	42268900	28623100

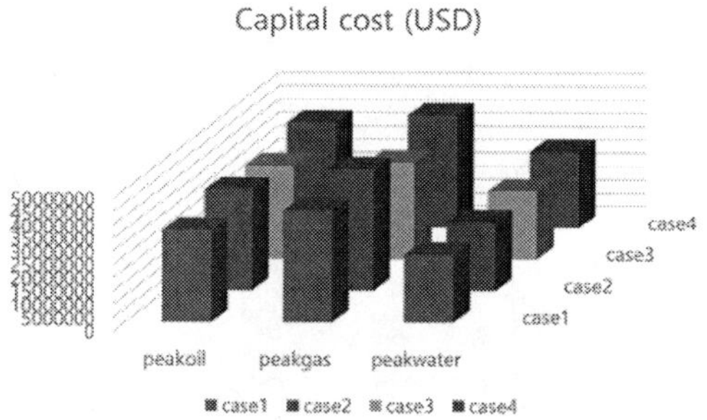

* Capital cost and installed cost were calculated by Aspen Process Economic Analyzer

5. Summary & Future Plans

4. Summary

Summary

- Case 3 and 4 have condensate recycle scheme to make the process produce more oil product. Heavy hydrocarbon components carried over to gas flow are recycled to oil flow and only light hydrocarbon components flow toward gas process.
  It means that separation efficiency of whole process was improved and hydrocarbon recovery was increased.

- As more oil flow enter the separator through recycling, duty of pre-heaters increased while power of compressors decreased. Total utility price is almost constant whether it has recycle.

- As separator is added, heat duty increases and electricity decreased. But Total utility price increases when the process has more separator.

- Comparison including capital cost, operating cost, product sales shows that Case 3 is the most economically feasible.

- Best operating condition was drawn through Simulation based optimization using GA .

MEETING ADDITIONAL PLANT CAPACITY REQUIREMENTS WITHOUT ADDING ADDITIONAL VESSELS

Gene Eberhardt
Woven Metal Products - Alvin, TX USA

Gene Eberhardt is the Vice President of the Reactor Division for Woven Metal Products in Alvin, Texas. He holds a Bachelor of Science Degree in Business from Saint Mary's University of Minnesota, a Master's Degree in Business Administration (MBA) from Purdue University and a Master's Degree in International Business Management from TiasNimbas Business School - The Netherlands. He holds a patent for Catalyst Bed Support systems and has several patents pending. He continues to develop industry changing technologies and methods in collaboration with Purdue University.

Given today's demanding throughput needs to maximize plant profits, plant de-bottlenecking quickly becomes an area ideal for identification and immediate return on investments. Today's gas plants require systems optimization to safely produce and deliver predictable, profitable products, longer predictable drying cycles and increased molecular sieve bed performance. These are not off-the-shelf solutions and require robust engineered solutions that are performance driven, industry proven and readily available for demanding plant outage requirements.

Molecular sieve bed support is one of, if not the most critical mechanical components within the dehydration vessel. Once the bed support requirements are understood, all options can be evaluated and a competent molecular sieve bed plan can be developed.

As with many natural gas plants commissioned over the years, molecular sieve bed support and proper molecular sieve bed utilization have been a secondary concern in these plant designs. In many cases, operational conditions have not allowed for these units to be properly evaluated or engineered to meet the operational needs of the plant. It has become an industry standard to install an outlet collector (elephant stool) in the outlet nozzle of the vessel with very little engineering emphasis given to the outlet collector, the effects that the outlet collector has on the molecular sieve bed, the performance of the molecular sieve bed or the effects the outlet collector has on the life of the molecular sieve bed. In many applications, the outlet collector serves only as a containment device and has not been engineered to aid in the bed performance or in the bed longevity. Many of the outlet collectors utilized today include an assortment of materials and configurations that include (but not limited to) slotted pipe, rolled plate, engineered support frames and a variety of material coverings including – wire mesh in a variety of material grades, wire mesh sizes and open area. In addition to the many wire mesh offerings many licensors and fabricators select profile wire in a wide range of stainless steel material grades and slot openings all of which do not observe any industry standards or a standard engineering design for these applications.

In some applications, a dehydration unit will have a horizontal bed support installed near the tangent line or the head seam of the vessel with a variety of materials used for bed containment including layers of wire mesh wire tied over grating and profile wire grid arrangements. In these bed configurations, the outlet collector is often installed as a secondary containment device as most vessels are not rolled perfectly round creating challenges to accomplish a full perimeter seal. The horizontal bed support configuration is a significant upgrade to the traditional outlet collector only configuration, however the molecular sieve bed design and perimeter sealing must be fully understood with emphasis on a secondary containment device to prevent total bed migration if the perimeter seal becomes compromised. The horizontal bed support option does offer much better bed utilization; however, this does sacrifice the total bed volume as the head of the vessel (at a minimum) becomes a void space.

This paper will focus on a recent installation of the Woven Metal Products / Haldor Topsoe co-developed (U.S. Patent Number 9409137) Outlet Support Grid (OSG) bed support system for the Blue Racer Midstream Company – Natrium West Virginia Complex, which consists of two cryogenic natural gas processing plants, each of which has 200 MMcf/d of processing capacity. This plant design includes 3 industry standard dehydration vessels that have an inner diameter of 84 inches (7 feet) and a total seam length (not including vessel heads) of 164 inches (13 feet 8 inches). When operational, this plant has been designed for parallel adsorption (Down Flow) and one vessel in regeneration (Up Flow) / standby.

Woven Metal Products Outlet Support Grid U.S. Patent 09409137

The Outlet Support Grid now offers an engineered total bed solution utilizing the performance of a horizontal bed support with the containment reliability that an outlet collector provides. Often the additional bed volumes realized with the Outlet Support Grid system installed will provide 15% or more additional bed volumes depending on the elevation of the horizontal bed configuration it replaces. As shown in Figure 1 the Computational Fluid Dynamics (CFD) modeling of the Outlet Support Grid shows a very uniform flow pattern which is critical for optimized bed performance, combined with full bed containment the Outlet Support Grid system is a reliable operational solution.

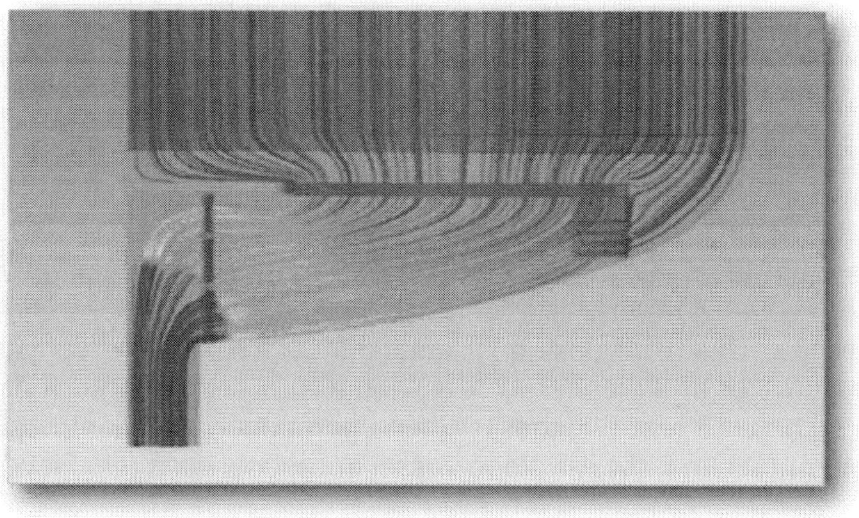

Figure 1.

History

This CB&I natural gas plant design was commissioned in April 2015, at which time operations had encountered / suspected a bed lifting (bed "fluffing") that resulted in an unusually high pressure reading across all three molecular sieve beds of 130 psi ΔP indicated. This unusually high ΔP was also accompanied with heavy dusting and plugging in the units down-stream filters, which prompted the operations team to suspect molecular sieve crushing to have incurred during the startup of the plant.

New molecular sieve was promptly ordered for the three dehydration vessels and the unit outage was taken for the replacement of the suspected crushed molecular sieve. During this outage, the unit beds were not fully emptied or cleaned only the molecular sieve was replaced and then the beds were put back in service with the existing bed support media undisturbed and no inspections or cleaning of the outlet baskets or inert media was performed. These three units were restarted and the high ΔP remained at 130 psi and continued to climb.

The Blue Racer operations team was now faced with the task of discovering any and all options to correct the high ΔP situation and restore the plant to a safe operational rate, quickly. The Blue Racer operations and engineering teams contacted their local UNIVAR field representative to source not only the new molecular sieve and new inert media support balls for each dehydration bed but they also started looking for replacement options for the potential of a semi crushed or fully crushed outlet basket in each of the three dehydration units. Woven Metal Products was contacted and was informed of the current operational situation and the challenges facing the Blue Racer team.

Collectively the Blue Racer, Woven Metal Products, UNIVAR, Zeochem and Catalyst Force teams addressed the following items; safety risks of continued operations, critical project timeline, operational limitations, current bed configuration, and equipment history of the installed equipment.

Each of the three dehydration vessels were originally equipped with outlet baskets sized at 15 inches outer diameter, 10 inches overall height with a profile wire covering, then welded to the outlet nozzle. These outlet baskets were constructed of 304 stainless steel, designed for 50 PSIG and 30% - 40% open area. Each dehydration vessel was loaded as follows:

Pre – Installation Bed Loading			
Media	**Pounds**	**Cubic Feet^3**	**Feet / Inches**
Ceramic Inert Balls (size unknown)	820 (estimate)	19	0.5 Feet
Guard Bed Media	1,653	37.57	0.98 Feet
4 x 8 mesh 1/8" Molecular Sieve	12,600	273.91	7.12 Feet
8 x 12 mesh 1/16" Molecular Sieve	7,200	153.19	3.98 Feet
Ceramic Inert Balls (size unknown)	820 (estimate)	19	0.5 Feet
Totals	23,093 (estimate)	502.67	13.08 Feet

The Blue Racer operational records reveal an average ΔP of 134 PSIG with a maximum ΔP of 180 PSIG indicated in table 1 (below) given the very high ΔP the actual condition of the outlet basket and molecular sieve bed was unknown. The Blue Racer operations reported a high frequency of heavy filter dusting / plugging, exchangers plugging, high system pressure requirements, dew point control issues, overall output product controls and longer bed regeneration times.

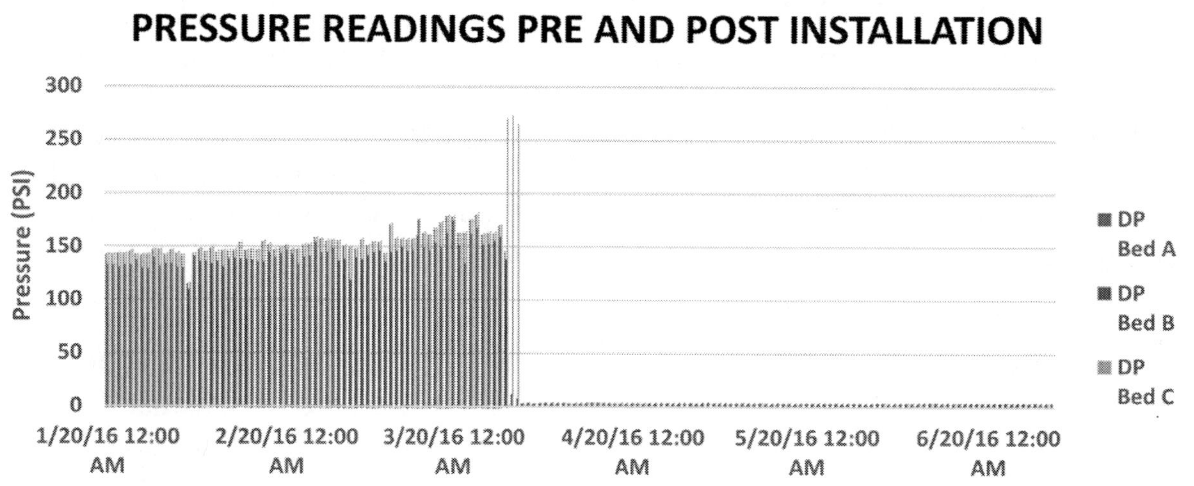

Table 1.

The following options were considered given the current vessel configuration and the critical path for both fabrication and the unit outage requirements;

- Replace the current outlet baskets with a design that would allow more open area and sized to the vessel.
- A horizontal support grid option was not considered due to the dehydration vessels not originally equipped with a support ring or beams and the time required to modify the vessels.
- Outlet Support Grid (OSG) system.

Given Blue Racer's operational experience with the outlet baskets the evaluation and decision was made to install the Outlet Support Grid solution. Woven Metal Products began engineering the OSG solution with fabrication drawings provide to the client for approvals in two weeks and full fabrication / delivery promised in just four weeks. The team confirmed that the Zeochem molecular sieve and bed support media would be onsite for the outage date and the Catalyst Force team would be able to service the outage with the required staff to execute a safe and timely outage. Schedules were drafted for the expedited fabrication and for the unit outage, both with very tight time lines. The following table (Table 2) shows the outage plan with just 26 hours allocated for a blind to blind per bed installation plan.

Event	Scheduled Time	Actual Time	Difference
Install Blinds	2.0 Hours	2.0 Hours	
N2 Blowdown	2.0 Hours	2.0 Hours	
Open Top Manway	1.0 Hours	0.75 Hours	0.25 Hours
Obtain Entry Permit / Safety review Meeting	1.0 Hours	1.0 Hours	
Equipment Rigging	2.0 Hours	2.0 Hours	
Unloading Bed	3.0 Hours	2.0 Hours	1.0 Hours
Vessel Cleaning	1.0 Hours	0.5 Hours	0.5 Hours
Remove Outlet Basket	2.0 Hours	0.5 Hours	1.5 Hours
Install OSG	4.0 Hours	1.5 Hours	2.5 Hours
Loading Molecular Sieve Bed	4.0 Hours	2.5 Hours	1.5 Hours
Close Manways / Un-install Blinds	4.0 Hours	3.5 Hours	0.5 Hours
Total Hours **Table 2.**	**26 Hours**	**18.25 Hours**	**7.75 Hours (Schedule Positive)**

The Outlet Support Grid is designed and installed without the need to weld to the vessel, no rope packing, no sealing materials and no thermal blankets are required to form a reliable sealing solution to the vessel head. The Outlet Support Grid mechanical sealing system preforms extremely well over the vessel and process thermal cycles which is critical in the full bed containment solution preventing media migration.

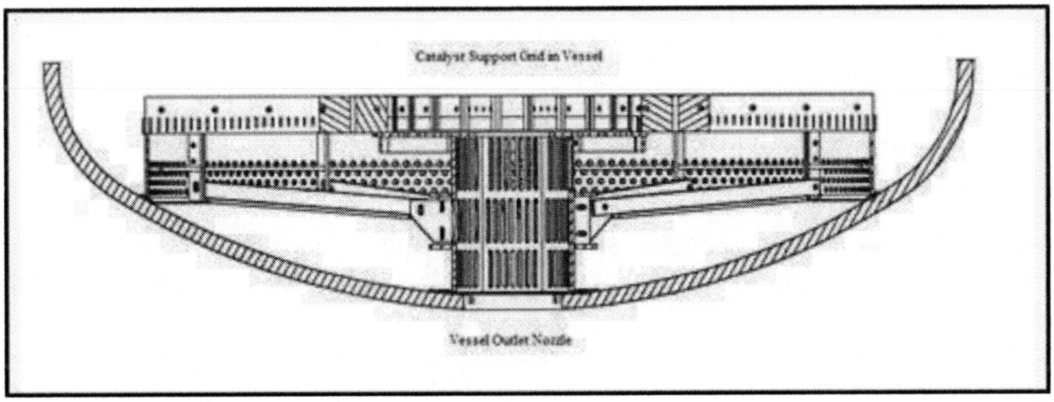

Installed View

Given the safety concerns of the high ΔP of 130 PSIG and rising, Woven Metal Products Outlet Support Grid was engineered to address these concerns and meet the bed design pressure drop of 3.08 PSIG at startup. With properly designed profile wire applied on all grid surfaces, the Outlet Support Grid has shown and will continue to perform at a very low and predictable ΔP as shown in table one since start-up. The Outlet Support Grid is designed to be in direct contact with the molecular sieve or any other media as the profile wire slot spacing is designed for the catalyst size loaded into the bed. Any support media utilized further assists in the bed hydraulics and should be designed and loaded for this purpose. As loaded the following inert balls were utilized 1/2" (bottom), ¼" (mid layer) and 1/8" (top layer) in contact with the molecular sieve, each application should be carefully examined and designed by catalyst providers.

Post – Installation Bed Loading			
Media	**Pounds**	**Cubic Feet³**	**Feet / Inches**
Ceramic Inert Balls ½"	828 (estimate)	19	0.5 Feet
Guard Bed Media	2,200	50	1.8 Feet
4 x 8 mesh 1/8" Molecular Sieve	10,500	228.26	6.1 Feet
8 x 12 mesh 1/16" Molecular Sieve	6,000	127.66	3.98 Feet
Ceramic Inert Balls 1/8" (Bottom – Top)	414 (estimate)	10	0.25 Feet
Ceramic Inert Balls ¼" (Bottom – Middle)	414 (estimate)	10	0.25 Feet
Ceramic Inert Balls ½" (Bottom – Bottom)	414 (estimate)	15	0.25 Feet
Totals	20,770 (estimate)	459.92	13.13 Feet

Blue Racer operations utilize a timer method for bed cycling; previous timer cycles were set at adsorption for 12 hours and regeneration for 6 hours. Post installation adsorption cycle is now set at 16 hours and regeneration is set at 6 hours. With the molecular sieve bed hydraulics optimized, the bed timers are being increased steadily over time while watching for potential bed break through. The initial bed performance indications and the plant records for these dehydration beds show minimal filter dusting after one year of continuous processing. These are all positive indications that these molecular sieve beds will meet or exceed the estimated 3 years of bed life giving the Blue Racer operations a much safer and very predicable operation.

This project and the installation were considered very successful with the dehydration vessel and the bed support concerns addressed and corrected including;

- Maximum volume of molecular sieve for the unit
- Longer molecular sieve bed adsorption cycles which enhance throughput / plant de-bottlenecking
- Fewer bed regeneration cycles reduces the amount of molecular sieve breakdown over time
- Full bed containment
- Full bed utilization
- Consistent dew point control
- Predictable bed performance
- Reduce fuel gas consumption

Outside the vessel benefits can be examined and appreciated with full bed utilization, fewer bed regeneration cycles will be required. The operational impact of fewer bed regeneration cycles is easily calculated in the reduction of valve cycles – given each bed includes 6 inlet /

outlet valves, incoming wet gas feed valves, steam and heater control valves essentially doubling the operational life of these valves greatly reducing the required maintenance costs for each valve. Each plant should consult with their valve manufacture / supplier to fully understand these operational valve benefits and the potential maintenance cost savings. Full bed containment aids not only in the valve life with less contamination present, full bed containment greatly reduces downstream filter plugging issues, cryogenic expander unit plugging and the need to bring the operational plant down, open and clean these units at un-scheduled times and loosing valuable production profits.

Conclusion

As with all plants, processes and capital projects many factors must be considered and evaluated. Return on investment (ROI) calculations can include many of the areas evaluated and discussed in this field installation and within this paper. Many factors will need to be evaluated including personnel and plant safety, natural gas prices, operational needs, operational costs, opportunity costs associated with lost production, lost efficiencies in molecular sieve beds and the installed equipment that are not engineered to maximize production capabilities and capacities. More information on the Outlet Support Grid and many other equipment solutions can be found at our website www.wovenmetal.com

Development of Novel Approach to Pretreat Natural Gas at Off-Shore Plants

2017. 03. 27

Soojin Kwon

Index

1. Introduction of natural gas pretreatment

1. Introduction - natural gas process configuration

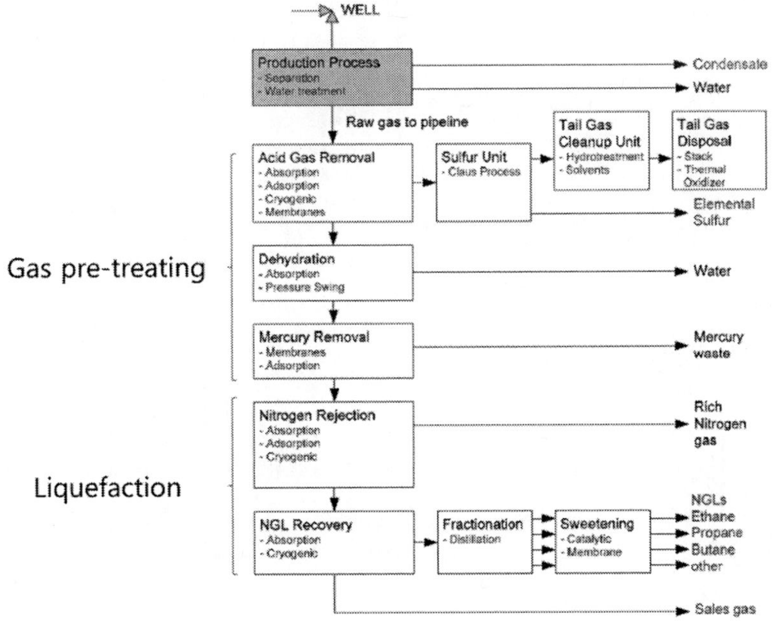

1. Introduction - Acid gas removal

- Absorption

Undesired compounds in the gas stream must be slightly soluble in the liquid, which is categorized as an absorbent.

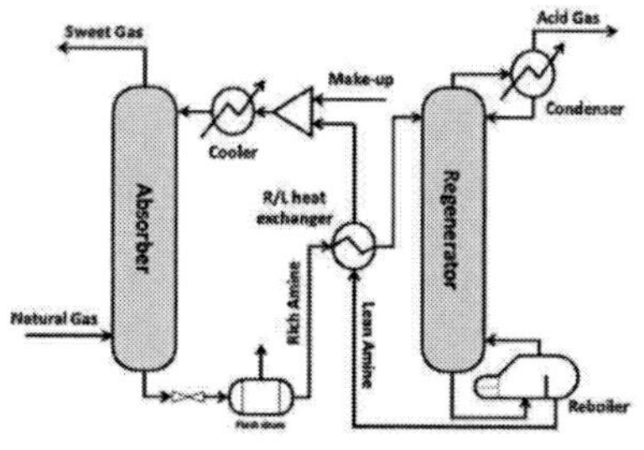

A typical amine-based acid gas removal unit.

- Membrane

- Energy saving
- Quite simple
  no mooving parts, compact, easy to scale-up, relatively easy to operate and control

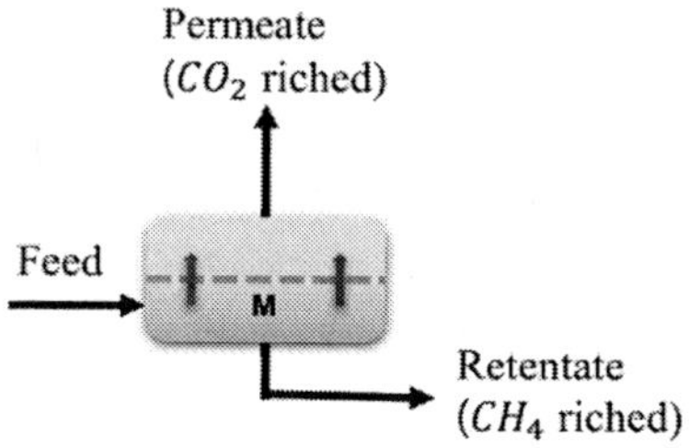

1. Introduction - various configuration for acid gas removal

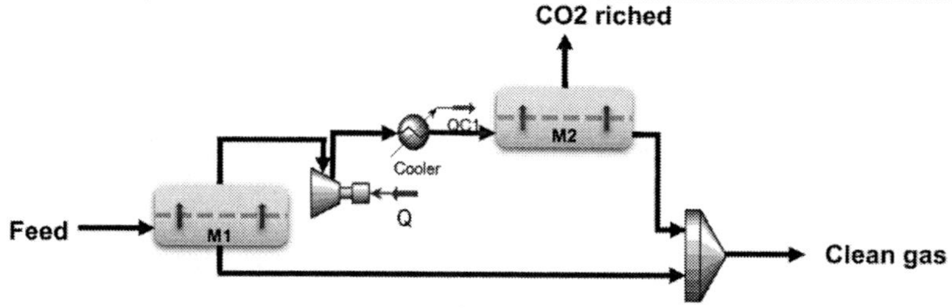

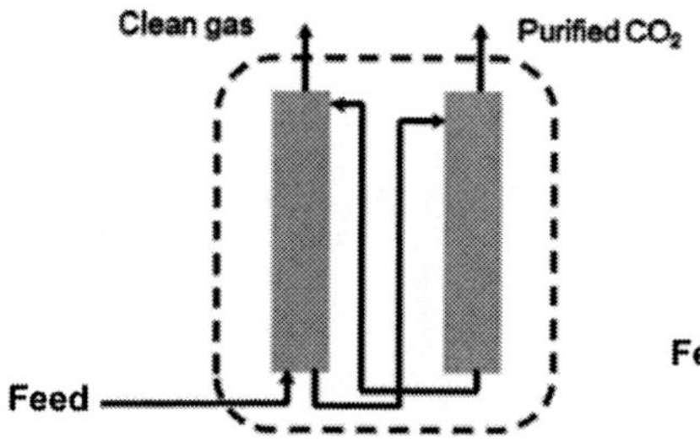

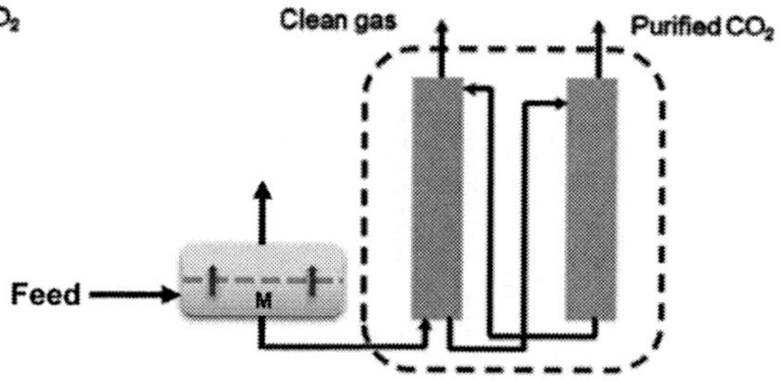

2. Research contents

2.1. Modeling of membrane system

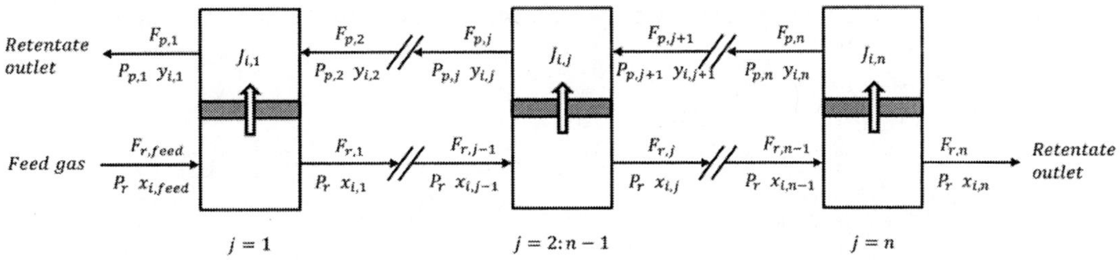

Schematic diagram of hollow fiber membrane separation.

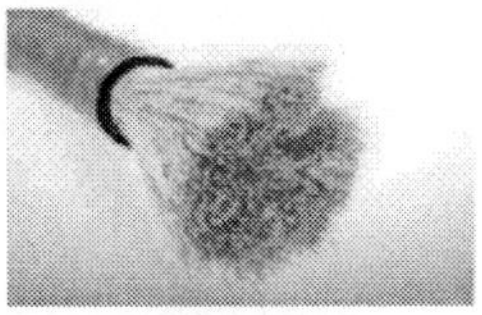

- Asymmetric cellulose acetate hollow fiber was modeled.
- This study considers counter-current configuration
- For the longitudinal flow configuration (considering variation of compositions and flow rates of both shell and tube sides along the axial direction), this model is divided into cells in the axial direction.

2.2. Case Study

Operating and feed conditions used in the membrane modeling

Component	Feed composition (mole fraction)	Operating conditions	
CO_2	0.172	Feed flow	$32.8 Nm^3/s$
CH_4	0.7759	Feed pressure	80 bar
C_2H_6	0.0417	Permeate pressure	4 bar
C_3H_8	0.0104	Flow pattern	counter-current

Membrane properties	
Inner diameter	200μm
Outer diameter	500μm
Effective length	3m
Product purity target	$x_{CO_2} \leq 2 mol\%$

Permeance and selectivity (Pan, 1986)

Permeance [10^{-8}mol/sm^{2}Pa]		Selectivity	
		CO_2/CH_4	36
CO_2	1.34	C_2H_6/CH_4	0.27
		C_3H_8/CH_4	0.054

Note : Feed condition data is extracted from gas reservoir that is located at west coast of Australia.

2.2. Case Study

Four configurations used to compare each methane recovery of product stream and economic evaluation.

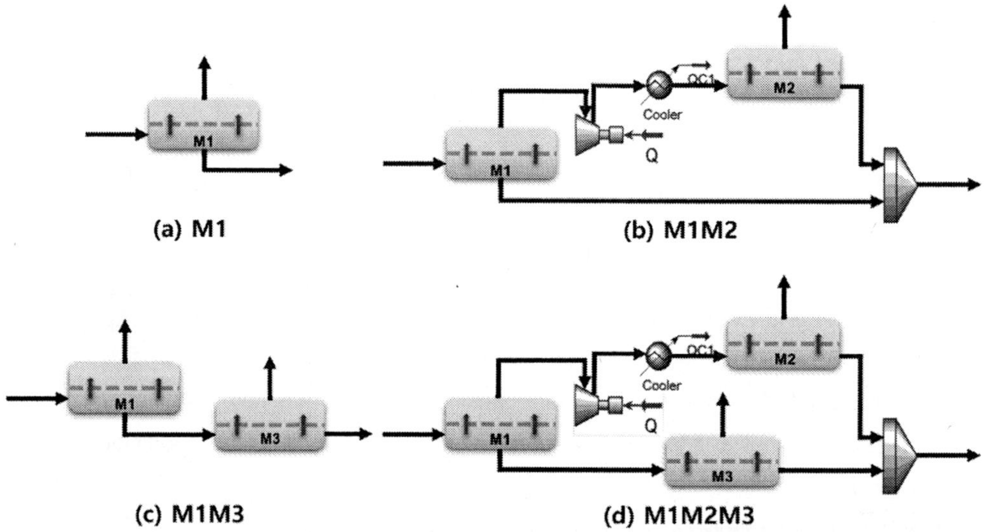

(a) M1 (b) M1M2 (c) M1M3 (d) M1M2M3

2.2. Case Study

Economic parameters and assumptions for computation of GPC (Hao et al., 2002)

Total Plant Investment	**TPI**	**TPI = TFI + SC**
Membrane module cost	MC	$ 10/ft2 (includes cost of membrane element)
Fixed cost	FC	MC + CC
Base Plant Cost	BPC	1.12 × FC (includes home office cost : 0.12FC)
Project Contingency	PC	0.2 × BPC
Total facilities investment	TFI	BPC + PC
Start up cost	SC	0.10 × VOM
Annual Variable Operating and Maintenance Cost	**VOM**	**VOM = CMC + LTI + DL + LOC + MRC + UC**
Contract and Material Maintenance Cost	CMC	0.05 × TFI
Local Taxes and insurance	LTI	0.015 × TFI
Direct Labor cost	DL	$ 15/h
Labor Overhead Cost	LOC	1.15 ×z DL
Membrane Replacement costs	MRC	$ 3/ft2 of membrane
Annual Cost of CH4 Lost in Permeate	**CH_4LS**	**CH_4LS = NGLS * NHV * NWP**
heating value of natural gas	NHV	1000 Btu/ft3
wellhead price of crude natural gas	NWP	$ 3.026 /MMBTU
Annual Natural Gas Lost	NGLS	NGLS = 365 × OSF × Qf × yp(CH4) × xf(CH4)
Gas Processing Cost	**GPC**	**GPC = (CRC + CH_4LS + VOM)/[365 × OSF × Qf × (1-SCE) × 100]**
Annual Capital Related Cost	CRC	0.2 × TPI
Membrane Life	t	4 years
Heating Value of Natural Gas	1000	
On stream factor	OSF	96%
stage-cut equivalent (P_N/F_N)	SCE	
net feed flow rate	FN	
net permeate flow rate	PN	

The costs of compressor and cooler with utilities refer to Aspen Process Economic Analyzer (APEA).

2.3 Results Analysis - (a) M1

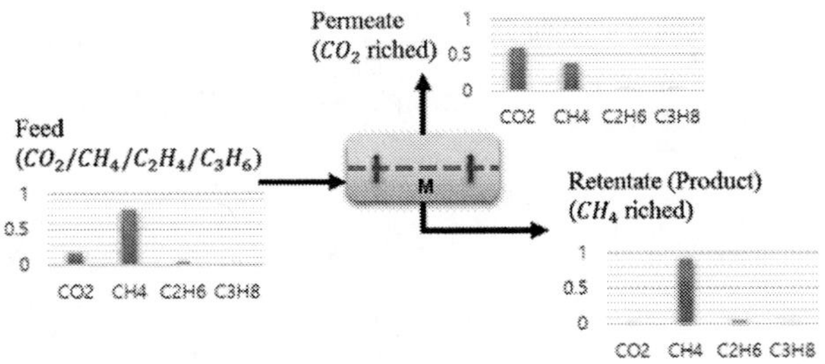

- CH_4 recovery : 87.1%
- ROI : 87.3%
- The number of fibers : 12,380,000
- Membrane area : 58,186 m^2

2.3 Results Analysis - (b) M1M2

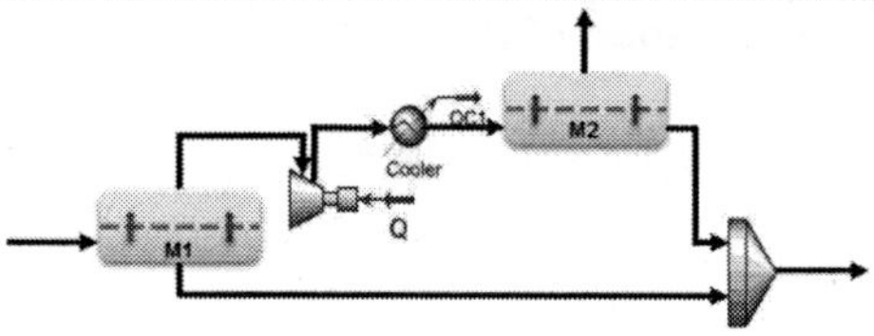

Methane recovery and ROI variation due to changes of the ratio of the membrane size of M2 membrane to that of M1 membrane (M2/M1).

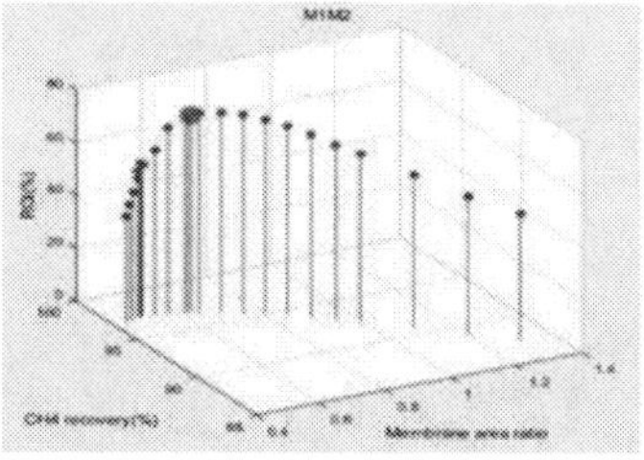

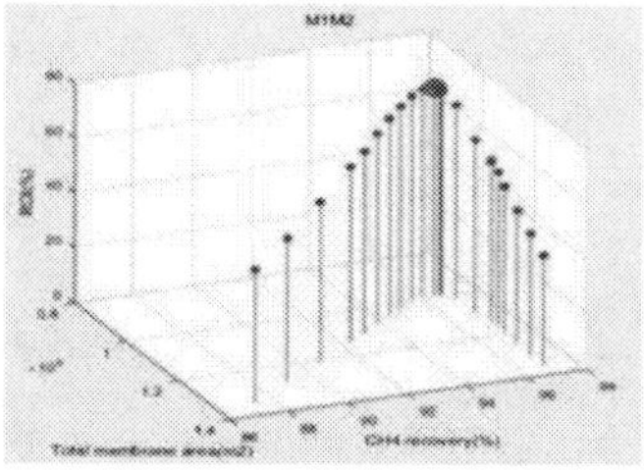

- A large amount of CO2 is mostly removed in M1, and the size of M2 membrane directly affects the methane recovery of product stream.
- If M2/M1 value is more than 1.4, the amount of retentate flow of M2 (Fr2) becomes negative (no flow).
- The optimum point that shows the highest value of methane recovery with the highest profit is confirmed from M1M2 configuration.

2.3 Results Analysis - (c) M1M3

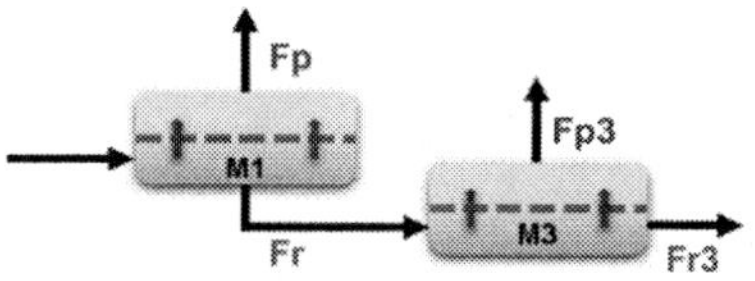

Methane recovery and ROI variation due to changes of the ratio of the membrane size of M3 membrane to that of M1 membrane (M3/M1).

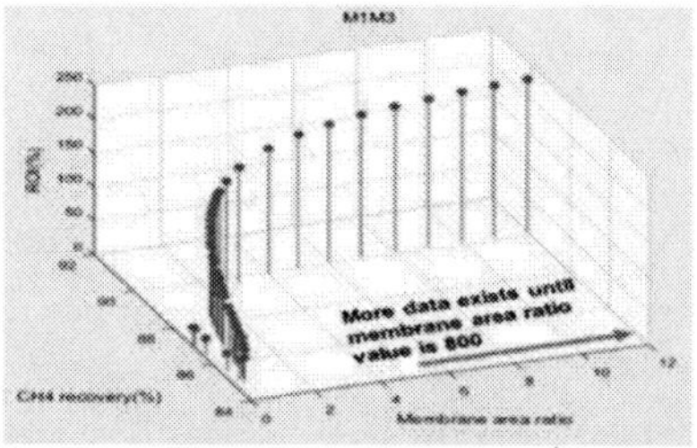

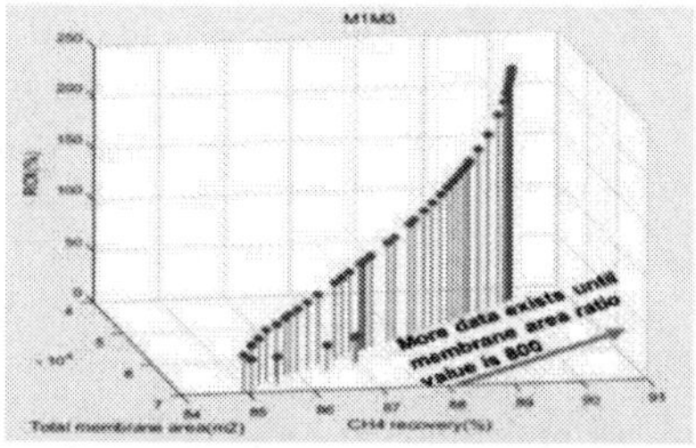

- If M3/M1 value is less than 0.05, the amount of permeate flow of M3 (Fp3l) becomes negative (no flow).
- If M3/M1 value is more than 10, the variation of ROI and methane recovery values can be considered negligible. That means, if M3 have a pretty big memabrane area in comparison with the area of M1, M3 acts like all alone.
- The total membrane area (M3+M1) become almost same especially over the point of M3/M1=10 until the amount of permeate flow (Fp1) reaches to the value of 0 (no flow) at about M3/M1=1000.

2.3 Results Analysis - (d) M1M2M3 (fixed M3/M1)

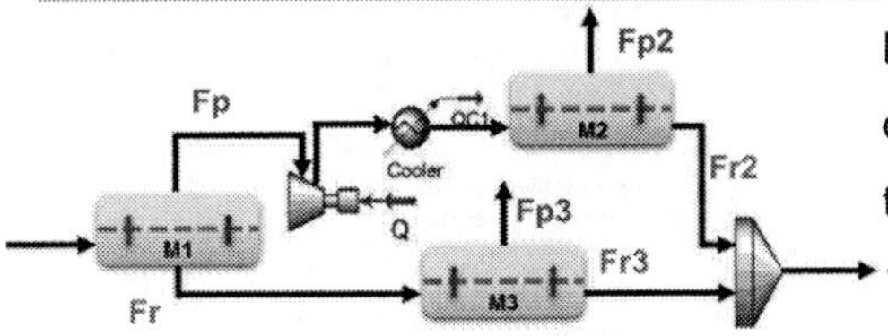

Methane recovery and ROI variation due to changes of the ratio of the membrane size of M2 membrane to that of M1 membrane (M2/M1) with fixed M3/M1 at 1.3 (the most reasonable value by trial and error).

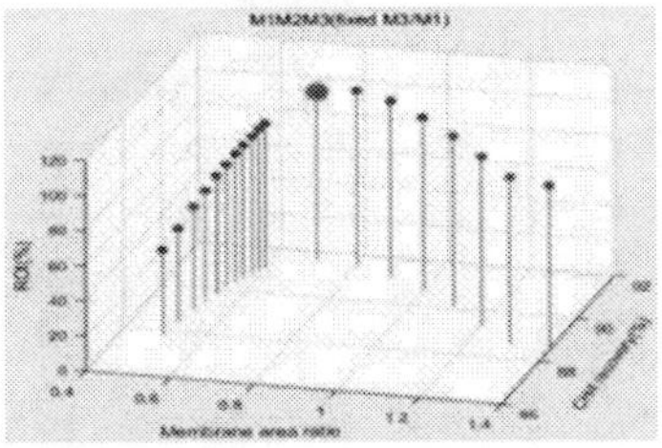

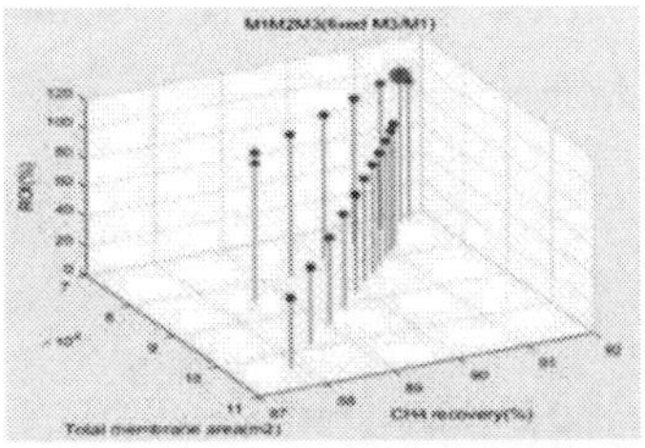

- If M2/M1 value is more than 1.5, the amount of retentate flow of M2 (Fr2) becomes negative (no flow).
- If M2/M1 value is less than 0.5, the membrane size of in M1 increases considerably in order to meet the pipeline specification of CO_2 concentration (<2mol%). It causes the total profit to become negative (no profit) because of the capital cost for membrane area.
- The optimum point that shows the highest value of methane recovery with the highest profit is confirmed in M1M2M3 (fixed M3/M1) configuration.

2.3 Results Analysis - (d) M1M2M3 (fixed M2/M1)

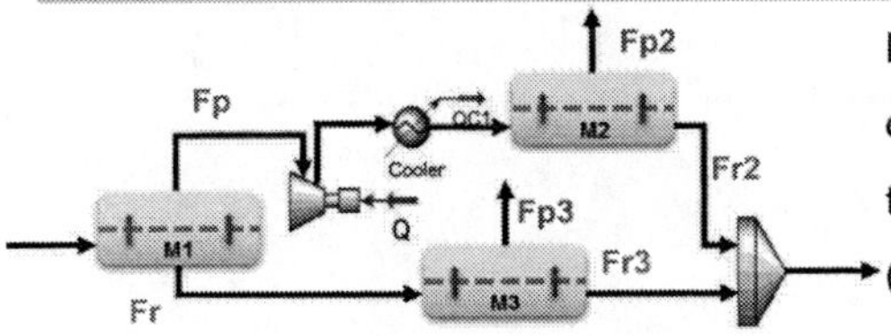

Methane recovery and ROI variation due to changes of the ratio of the membrane size of M3 membrane to that of M1 membrane (M2/M1) with fixed M3/M1 at 0.5 (the most reasonable value by trial and error).

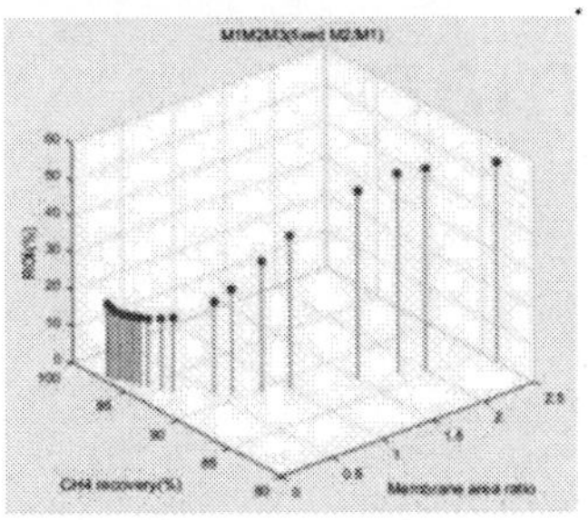

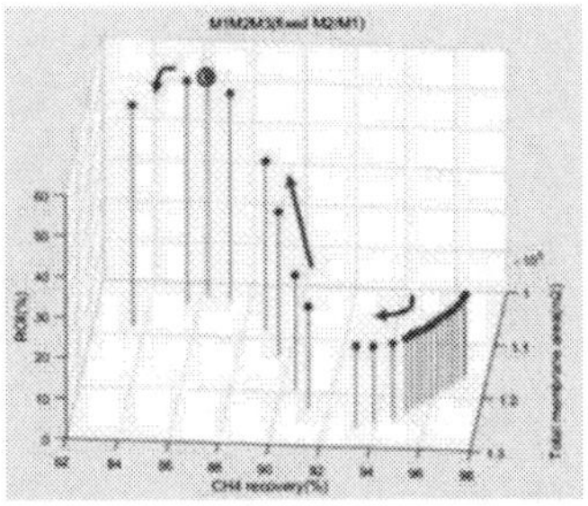

- If M3/M1 value is less than 0.05, the amount of permeate flow of M3 (Fp3l) becomes negative (no flow).
- M1M2M3 (fixed M2/M1) configuration has S-shape (middle section of graph shows inverse relationship between CH_4 recovery and ROI) that shows higher profit with lower methane recovery.

2.3 Results Analysis - (d) M1M2M3 (fixed M2/M1)

Membrane area in M1 decreases (**M1↓**) ⟹ Less permeate flow in M1 (**Fp↓**), More retentate flow in M1 (**Fr↑**)

Membrane area in M3 increases (**M3↑**) ⟹ More permeate flow in M3 (**Fp3↑**), Less retentate flow in M3 (**Fr3↓**)

→ Less product flow = Less methane recovery = **Less product profit**

<

Decreased total membrane area in both M1 and M2, **-Δ(M1+M2)**

V

Increased membrane area in M3, **Δ(M3)**

→ Less membrane area in total membranes **(M1+M2+M3 ↓)** = **Less capital cost**

Total profit increases (ROI ↑)

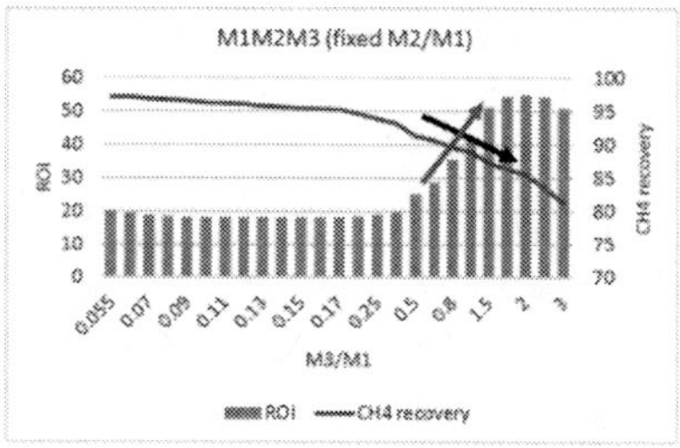

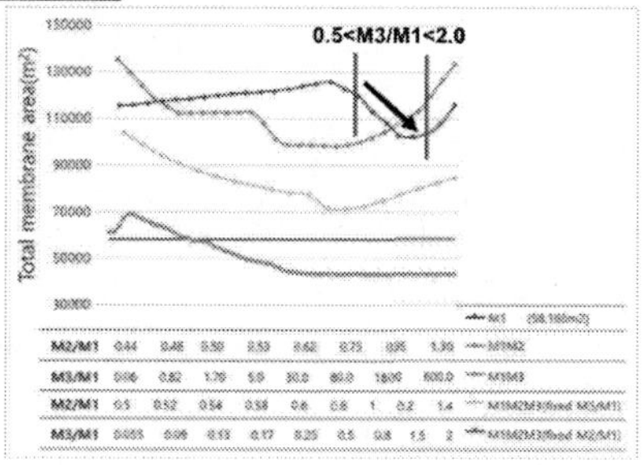

3. Summary

3. Summary

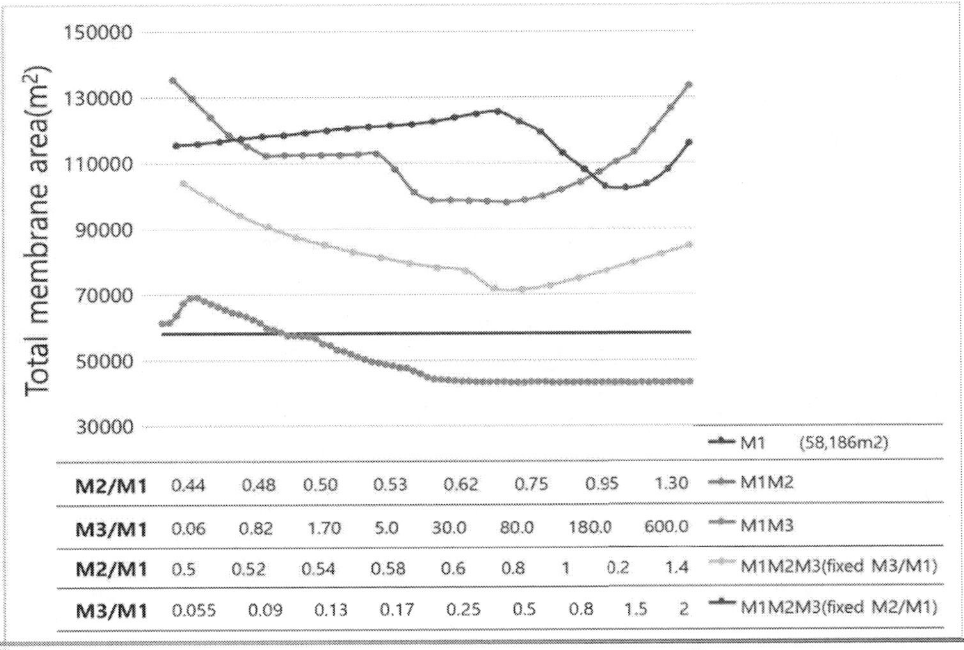

3. Summary

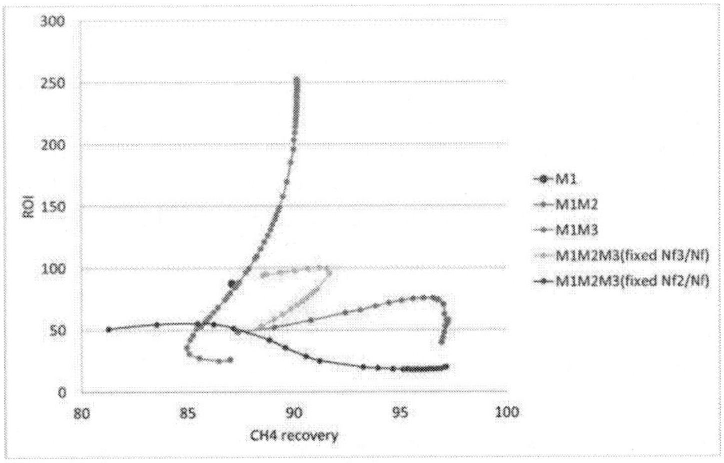

It provides the solution for the optimization of membrane configurations to maximize economic profit depending on required methane recovery meeting pipeline specification such as minimum CO2 concentration.

4. Future works

4. Future works

- Comparison economic analysis of membrane configurations with amine based absorption process including hybrid configuration.
- Continuous development of various configurations for natural gas treatment.

484898 Modularity and the Role of Process Intensification

Abstract:

The recent decades have witnessed significant efforts in both modular production and process intensification. As a common denominator, both process intensification and modular production are predicated on smaller, more efficient plants, where capacity is deployed as needed, and expanded by numbering-up rather than by conventional scale-up approaches. Modularity promotes agility and allows industrial players to enter markets and deliver products faster, with modular plants competing favorably with conventional facilities for certain types of products and market circumstances. In this talk, we discuss modular production in two contexts, centralized and geographically distributed manufacturing, and the impact of the choice of conventional or intensified technology on the economics of modular processes. We provide several examples and identify research needs in the design, control and operation of modular and intensified processes.

482645 Modular Manufacturing: Overcoming Cost and Reliability Barriers for Intensified Chemical Process Equipment

Abstract:

A growing trend in the chemical process industry for reducing capital costs and accelerating chemical plant builds involves building plants on site by assembling together modules, consisting of one or more unit operations, produced at a centralized manufacturing facility. The use of intensified chemical process equipment (ICPE), for reducing the size and weight of unit operations, is consistent with modular manufacturing needs to reduce the costs of producing, shipping and installing modules. However, the cost and reliability of ICPE technologies remains a barrier to the use of these technologies by module manufacturers. Recent manufacturing cost studies show that ICPE heat exchanger technology can be cost competitive with conventional technologies. Further, computational methods are being used to anticipate and overcome reliability challenges. Finally, new manufacturing process technologies will be discussed for further reducing the cost of ICPE.

482649 Process Intensification: AIChE *Rapid* Opportunities in Energy and Chemicals

Abstract:

Challenges in future energy and chemicals include expanded use of natural gas as a feedstock, incorporation of solar and wind into the energy supply, and enhanced concerns over emissions and footprints including land and resource utilization. The newly created ***Rapid*** institute for process intensification (PI) will be a game changer for AIChE in enabling the multi-disciplined collaborations needed to drive sustainable new technologies to commercialization, uniting efforts and capabilities from industry, academia, and national labs. Intensified and/or coupled reactions and separations will drive new paradigms for chemical engineering design to meet cost and efficiency targets. This seminar will provide a look back at both successes and challenges in the implementation of PI to date, and current opportunities for ***Rapid*** in meeting the very significant challenges we face in providing for a sustainable future supply of energy and chemicals.

PROCESS INTENSIFICATION IN THE B&P LITTLEFORD PLOW MIXER

Michael A. Smith PE, B&P Littleford LLC, Saginaw, MI

Introduction

Transfer of intermediates from one unit operation to the next takes time, space, and energy—and can be hazardous. Batch processes may be developed and designed to minimize cycle time, capital expense, and operating expense by incorporating process equipment designed to handle multiple unit operations in series. Horizontal plow mixers, such as the FKM and DVT series from B&P Littleford LLC, can handle operations from dry blending and granulation to interface chemical reactions and vacuum drying of dough-like materials.

Several processes are reviewed, each with multiple operations performed sequentially in a Littleford plow mixer. The key capabilities of the Littleford plow mixer are outlined for each unit operation. The impact on capital expense, operating expense, and cycle time for each process is also discussed.

Key Capabilities

The B&P Littleford plow mixer has several design features which enable it to be used in a variety of unit operations. Combined in various ways, these features allow for the development of multi-step processes while minimizing product transfers.

Heat Transfer

Most reaction vessels and distillation units are jacketed, with a heat-transfer fluid such as hot oil, chilled glycol or steam providing heat transfer. The close interaction of the plows with the wall in a Littleford mixer provides a fresh surface for heat transfer throughout the vessel, increasing the overall heat-transfer coefficient for unit operations such as drying.

Full Vacuum to High Pressure Capability

All models of B&P Littleford Mixers are capable of full vacuum. The DVT series adds double mechanical seals and pressure ratings of up to 300 psig for increased flexibility. Steam treatment of materials and pressurized reactions in solvents above their normal boiling point are made possible. Hazardous gases such as anhydrous HCl can be contained and safely reacted.

Granulation

The plow mixer is typically outfitted with one or more high-speed, high-shear "choppers" which act to break up agglomerates into smaller particles. The clearances between the chopper blades, the plows, and the vessel wall provide some limits to this action, along with moderating the speed and duration of the chopping action. Fine powders can often be granulated with the addition of a small amount of liquid, thorough mixing, and limited use of the choppers, resulting in a less-dusty product.

Mechanically-Fluidized Bed

At full speed, the plows are designed to maintain solid particles in a fluidized state, without gas flow or vibratory action. This allows for rapid and complete blending of solids, the efficient distribution of sprayed-in liquid additives, and uniform surface treatment of particles.

Peripheral Equipment

Vacuum systems, dust collection, metered addition systems and feed tanks can all be configured with the plow mixer in different ways, as needed for the particular trial.

Case Studies

The following case studies are actual tests conducted for customers. Confidential information has been withheld or permission given to share results. Each incorporates several unit operations without transferring product from one piece of equipment to another.

Vacuum Drying and Granulation

Many materials are formulated as a stable suspension or slurry, but are preferred to be shipped and handled as a solid. The preferred particle size may vary widely, but a "non-dusty" product presents fewer industrial hygiene concerns, and so a coarser granulation is desirable.

A typical application is seen in Figure 1. This product was temperature-sensitive, and therefore was dried under vacuum. As with many slurries which are dried to a powder, this material went through a cohesive "dough-like" phase, which required significant horsepower. The end of the cohesive phase is marked by a significant reduction in power draw. At this point the chopper was used to break up the product, which could then be more quickly and uniformly dried to the appropriate moisture. Granulation occurred at the same time, as large particles were broken down by the interaction of the chopper with the plow, and fine particles re-agglomerated to a characteristic size, as long as sufficient moisture remained.

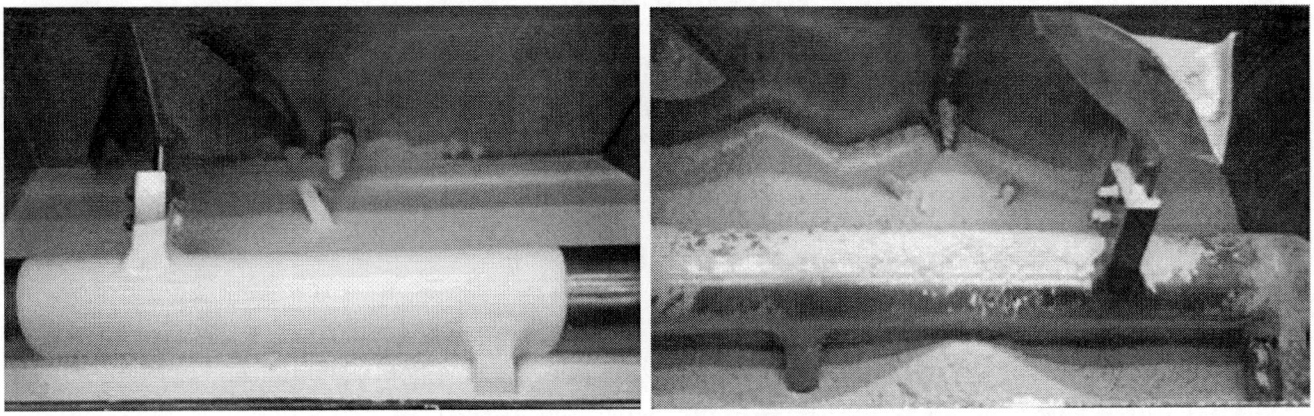

Figure 1: Product before and after vacuum drying and granulation

Limited-Water Reaction

Reaction of a phosphonate acid with ammonium carbonate, to produce the corresponding ammonium phosphonate, can be easily accomplished in a slurry.

$$2\ H_2PO_3\text{-}R\text{-}CO_2H + 3\ (NH_4)_2CO_3 \rightarrow 2\ (NH_4)_2PO_3\text{-}R\text{-}CO_2NH_4 + 3\ CO_2 + 3\ H_2O$$

Since the reaction generates water, only a small quantity of water is actually required to initiate the reaction. A significant reduction in reaction volume can be realized by reacting the two solids as a "wetcake" as long as the wetted solids can be vigorously mixed. Liberation of carbon dioxide gas makes this a net endothermic reaction.

The plow mixer provided the appropriate agitation and heat transfer to convert the acid to the salt quickly while maintaining the mixture at 17 to 20°C. Full vacuum and a 38°C jacket temperature was used to reduce the moisture to less than 6%. Finally a second ammonium additive was blended into the product before discharge. Figure 2 depicts a process flow chart, which shows the five separate unit operations which were performed in sequence, in a single vessel, without transferring the material.

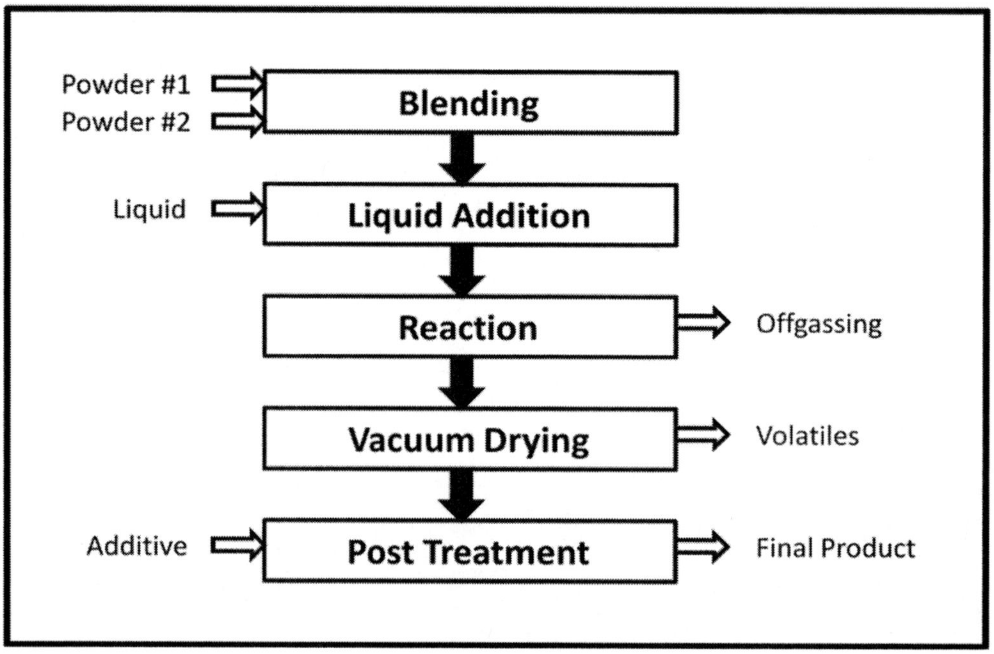

Figure 2: Typical Process Flow

Polymer Spray-Drying

A methacrylate copolymer dissolved in a flammable solvent was pre-heated and sprayed into the plow mixer, which was heated and under vacuum. The flashing solvent was condensed and recovered via the vacuum system, while the polymer formed a fine powder. Once formed, the powder remained fluidized by the action of the plows. This powder, heated by the walls, served as a substrate for additional sprayed product. Particle size was controlled by use of the chopper to break down any agglomerates that formed.

Surface Treatment of Clay

Certain minerals are found to have increased utility when surface-treated with chemical modifiers, such as a quaternary ammonium salt. Clays present a particular problem, in that they tend to become extremely cohesive when wetted. This results in more shear-induced heating, which further increases the cohesion.

In this example, a powdered clay material was first pH-modified with caustic and water, and heated to the desired temperature for reaction. The mixture was agitated at low speed. The quaternary ammonium compound was then metered in over several minutes, while the power draw of the main agitator was monitored. After observing the amperage peak, the agitator speed was increased and the choppers used, while applying vacuum to dry and granulate the product.

Chemical Vapor Deposition

Chemical vapor deposition (CVD) was accomplished, using the mixer as a mechanically fluidized bed. The substrate was first pre-heated and dried under high vacuum. With the substrate fully fluidized, vacuum was used to introduce the volatile reactants into the vapor space, alternating from one to the other in order to build up the desired surface layer.

Direct-Contact CO_2 Chilling

A food customer required rapid blending of dry ingredients with a low-melting oil, resulting in a free-flowing powder. The dry ingredients were pre-chilled with the direct injection of solid CO_2 powder via a "snow horn." With full fluidization, the warmed oil was sprayed in, followed immediately with additional solid CO_2, which quickly cooled the material to the desired packaging temperature.

Discussion

As can be seen by the above case studies, the B&P Littleford plow mixer is a useful and highly configurable processing unit. It is ideal for a variety of unit operations, and at least capable in many others. As such, it can be used to develop "one-pot" processes for the economic scale-up of new batch processes. One multi-purpose piece of equipment will require less capital than several one-use machines, which may or may not be repurposed.

At the Market Development stage, an appropriately sized plow mixer may be tasked to perform the entire process initially. It can also be used to produce sufficient intermediates for use in other equipment testing, for example if a traditional spray dryer might be more economical for the drying step, or a ribbon blender for a final additive and packaging hopper.

From a Process Safety Management perspective, the fewer product transfers done with (potentially) hazardous new products and mixtures, the better. This minimizes the potential for personnel exposure as well as flammability concerns, environmental release, etc. This is true at every stage of process development, but especially in the pilot plant stage, when the volume of chemicals becomes significant, but experience with the process is still relatively low.

481219 Thermodynamic Modeling of Aqueous Nickel Nitrate-Nickel Chloride-Nickel Sulfate System with Electrolytes NRTL Model

Abstract:

Thermodynamic Modeling of Aqueous Nickel Nitrate-Nickel Chloride-Nickel Sulfate System with Electrolytes NRTL Model

Yuan Li, Benjamin Caudle, Chau-Chyun Chen

Texas Tech University

Abstract

Millions of tons of electronic waste such as computers and mobile phones, known as e-waste, are produced each year. Traditionally most e-waste has been disposed of in landfills [1], but in recent years recycling of e-waste is being taken seriously due to environmental concerns and potential financial gains from recovery of materials. In one mode of recycling, metals in e-waste are dissolved in strong acids and then the valuable metals are recovered through electrochemical methods. Nickel is one of the significant metals in e-waste and its dissolution and saturation can significantly affect the efficiency of the recovery system. Therefore, a comprehensive and validated thermodynamic model for nickel salts in strong acids is a necessity for process simulation of e-waste recycling processes.

With a wide applicability from dilute to saturated electrolyte solutions, the electrolytes NRTL (eNRTL) model is one of the most accurate thermodynamic models for aqueous and mixed solvent electrolytes [2-4]. Moreover, the eNRTL model has been successfully validated for various strong acids and mixed acid systems [5-7]. In this work, we extend the eNRTL model and develop a comprehensive thermodynamic model for nickel salts in strong acids including nitric acid, hydrochloric acid, and sulfuric acid.

We first identify the eNRTL binary interaction parameters for aqueous nickel nitrate, nickel chloride, and nickel sulfate electrolyte systems from available literature data using phase equilibrium properties such as vapor pressure, osmotic coefficient, mean ionic activity coefficient, and calorimetric properties such as molar heat capacity and excess enthalpy. Salt solubility data are then used to identify thermodynamic properties for corresponding salt crystals. The binary system models are then integrated with the strong acid models developed with eNRTL [5-7] and further validated against available data for ternary nickel salt - acid – water systems. These ternary system models will form a sound thermodynamic foundation for use in process simulation of e-waste recycle processes.

[1] Robinson, B.H., E-waste: An assessment of global production and environmental impacts. Science of the Total Environment 408 (2009) 183-191.

[2] Chen, C.-C. *et al.*, Local Composition Model for Excess Gibbs Energy of Electrolyte System. Part I: Single Solvent, Single Completely Dissociated Electrolyte Systems. AIChE Journal 28 (1982) 588-596.

[3] Chen, C.-C. *et al.*, A Local Composition Model for the Excess Gibbs Energy of Aqueous Electrolyte Systems. AIChE Journal 32 (1986) 444-454.

[4] Song, Y. and Chen, C.-C., Symmetric Electrolyte Nonrandom Two-Liquid Activity Coefficient Model. Ind. Eng. Chem. Res. 48 (2009) 7788-7797.

[5] Wang, M. *et al.*, Thermodynamic Representation of Aqueous Sodium nitrate and Nitric Acid Solution with Electrolyte NRTL Model. Fluid Phase Equilibria. 407 (2016) 105-116

[6] Hassanjani, S. et al., Comprehensive Thermodynamic Modeling of Complex Mixed-Solvent Electrolyte Systems: An Investigation on Water-Hydrogen Chloride-Methanol Ternary System. In preparation.

[7] Que, H. *et al.*, Thermodynamic Modeling of the Sulfuric Acid - Water - Sulfur Trioxide System with the Symmetric Electrolyte NRTL Model. J. Chem. Eng. Data. 56 (2011) 963-977.

481220 Thermodynamic Modeling of Ion Adsorption in Capacitive Deionization Cell Units with Enrtl Model

Abstract:

Capacitive deionization (CDI) is a technology to remove salts from low salinity brine solution with electrical cell units. The two electrodes in each cell unit adsorb opposite charged ions in the brine water when charged and release them when discharged. Modified Donnan's theory has been used to model the electrical double layer surrounding electrodes. However, non-ideality of ions adsorbed to the electrodes has yet to be adequately accounted for. [1]

Our previous work established a thermodynamic model for polyelectrolyte solutions by extending Manning's limiting law with eNRTL model representing short range interactions. For a polyion that has a charge density larger than a certain critical value, the counterions in the polyion solution are condensed on the surface of the polyion to reduce the charge density. Manning first described the limiting law for calculating counterion activity coefficients in polyelectrolyte solutions [2]. Our model extended the concentration range of salts so that it offers a comprehensive thermodynamic formulation to correlate and extrapolate thermodynamic behavior of all aqueous polyelectrolyte systems, including ion exchange membranes or resins.

The CDI units at equilibrium have the electrodes with counterions electrically attached to the surfaces. The scenario is very similar to the counterion condensation process of polyions. We apply the extended Manning's model to explain the amount of adsorbed salt in a CDI unit. We consider the electrodes as polyions. The polyions and adsorbed ions form a separate phase separated from the bulk brine solution. This polyions phase is then treated as the polyelectrolyte-water binary.

The activity coefficient of counterions on electrodes is calculated with the extended Manning model. From the equality of chemical potential of solute NaCl in the bulk solution and in the electrode polyion phase, we relate concentrations of bulk solution, amount of adsorbed salt, and charge density of the electrodes. Both the sodium cation and chloride anion are considered as counterions because the two electrodes attract oppositely charged ions at the two ends.

Predictions of salt concentrations in the electrodes at different bulk solution concentrations are generated and compared with experimental data. Future works include investigating the relationship between the model parameters and the cell voltage and finding optimal working conditions for a given desired separation.

Reference

1. P. M. Biesheuvel, et al., "Attractive forces in microporous carbon electrodes for capacitive deionization." *Journal of solid state electrochemistry* 18.5 (2014): 1365-1376.
2. Gerald S. Manning, "Limiting laws and counterion condensation in polyelectrolyte solutions I. Colligative properties." *The journal of chemical Physics* 51.3 (1969): 924-933.

481456 Thermodynamic Modeling of the Aqueous Copper(II) - Nitrate - Chloride - Sulfate System with the Electrolyte-NRTL Model

Abstract:

Thermodynamic Modeling of the Aqueous Copper(II) - Nitrate - Chloride - Sulfate System with the Electrolyte-NRTL Model

Benjamin Caudle, Yuan Li, Chau-Chyun Chen

Texas Tech University

Abstract

Electronic waste (e-waste) is waste associated with computers and cell phones, as well as increasingly digitized appliances such as refrigerators and ovens. As demand grows in emerging markets, the need for proper handling of this waste increases as well [1]. A unique characteristic of e-waste is the presence of both valuable materials, such as copper and members of the platinum group, and hazardous materials. Proper recycling of e-waste is necessary to prevent the release of toxic substances into the environment and has the added benefit of yielding materials that can be reused or sold. The most promising method of recycling is through leaching with mixed acids, in which metals can be selectively dissolved depending on the composition of the leaching solution. Copper is typically the most prevalent metal in e-waste and, even if it is not recovered for its own sake, its behavior in solution has a large impact on the system as a whole. For this purpose, a thermodynamic model of copper in common acid solutions is a useful tool for the design of recycling systems.

The electrolyte non-random two-liquid (eNRTL) model has proven to work for a wide range of electrolytes in both single and mixed solvents [2-4]. The nitric, chloric, and sulfuric acid systems already have validated eNRTL models available [5-7], and the prime advantage of the NRTL model is straightforward and accurate extension to other multicomponent systems.

This work extends the eNRTL model to cover the behavior of copper(II) in nitric, chloric, and sulfuric acids and mixtures thereof. Beginning with the binary systems, parameters are regressed from the available literature data. This includes phase equilibrium and calorimetric properties as well as solubility data for the respective copper salts. The binary models are then integrated with the existing acid system models and validated against literature data for the ternary and quaternary systems. The resulting model is an integral part of the overall model for e-waste recycling processes.

[1] Widmer, Rolf *et al.*, Global perspectives on e-waste, Environmental Impact Assessment Review, Volume 25, Issue 5, July 2005, 436-458

[2] Chen, C.-C. *et al.*, Local Composition Model for Excess Gibbs Energy of Electrolyte System. Part I: Single Solvent, Single Completely Dissociated Electrolyte Systems. AIChE Journal 28 (1982) 588-596.

[3] Chen, C.-C. *et al.*, A Local Composition Model for the Excess Gibbs Energy of Aqueous Electrolyte Systems. AIChE Journal 32 (1986) 444-454.

[4] Song, Y. and Chen, C.-C., Symmetric Electrolyte Nonrandom Two-Liquid Activity Coefficient Model. Ind. Eng. Chem. Res. 48 (2009) 7788-7797.

[5] Wang, M. *et al.*, Thermodynamic Representation of Aqueous Sodium nitrate and Nitric Acid Solution with Electrolyte NRTL Model. Fluid Phase Equilibria. 407 (2016) 105-116

[6] Hassanjani, S. et al., Comprehensive Thermodynamic Modeling of Complex Mixed-Solvent Electrolyte Systems: An Investigation on Water-Hydrogen Chloride-Methanol Ternary System. In preparation.

[7] Que, H. *et al.*, Thermodynamic Modeling of the Sulfuric Acid - Water - Sulfur Trioxide System with the Symmetric Electrolyte NRTL Model. J. Chem. Eng. Data. 56 (2011) 963-977.

479968 Careers in Process Development: How Aiche Can Help!

Abstract:

Process development is a career path with high visibility in industry, connecting idea generation through development to commercial deployment. The process development lead must act as an 'intrapreneur" to address all aspects of technology development, from process technology, to footprint and sustainability, and including process economics. This poster will provide an overview of Process Development activities, and how AIChE involvement can provide contacts and skills needed to excel. Also, how can AIChE better serve your needs?

480257 Optimal Design of Mixed Refrigerant System in Ethylene Plants for Maximum Energy Savings

Abstract:

Refrigeration system is one of the most important and critical operating system in the chemical and petrochemical industries. The tradition cascade refrigeration system (CRS) has been used for decades in ethylene plants. However, the use of the multiple refrigerants in the CRS requires a very number of separate compressors, heat exchangers, pumps and associated pipes for each temperature and pressure level. Mixed refrigerant system (MRS) has been introduced for decades, where the pure working fluids are replaced by mixed working fluids. MRS is more efficient in energy saving and less capital cost than CRS and has been widely applied in liquefaction of natural. However, because of the complexities of ethylene manufacturing process and refrigerant systems aforementioned as well as possible intellectual copyright concerns, the in-depth studies for a complex mixed refrigeration system design and operation are still lack of reporting. In this study, a general methodology has been developed for synthesis of MRS with a tertiary refrigerant, a mixture of methane, ethylene and propylene, to replace the ethylene-propylene cycles of the CRS in an ethylene plant. First, the exergy-temperature chart combined with the exergy analysis is presented to comprehensively analyze the thermodynamic nature of traditional CRS to help to locate and evaluate exergy loss in the system. Opportunities to improve are explored and identified in the exergy-temperature chart. Then, a superstructure is built from the conclusions of thermodynamic analysis to cover all the possible arrangements for the cooling supplies. Moreover, a simulation-assisted mixed-integer linear programming (MILP) model is developed and solved for the optimal process synthesis and heat exchange network (HEN) design. The case study about a mixed refrigerant system used in a typical ethylene plant is employed to demonstrate the efficacy of the developed methodology. The shaft work of compressor has been reduced tremendously. And the capital cost and utilities cost have been saved by 32% and 16%, respectively.

480957 Design and Development of a Supercritical Fluid Processing System for Conversion of Used Cooking Oil to Biodiesel

Abstract:

The design project involves collection of waste cooking oil from the University of South Florida and vicinity, conversion of waste cooking oil using a novel catalyst-free supercritical fluid processing facility, making novelty soap from the by-product glycerin, and utilizing the biodiesel in powering the University bus system that utilizes B20. The processing time is 15 minutes and allows a mobile solar powered unit. The design involves scale-up, heat integration, safety and environmental impact analysis.

2017 AIChE®

SPRING MEETING & 13TH GLOBAL CONGRESS ON PROCESS SAFETY

March 26–30, 2017
Henry B. Gonzalez
Convention Center
San Antonio, TX

17 AIChE Spring Meeting, San Antonio, TX

13TH GLOBAL CONGRESS ON PROCESS SAFETY

17 AIChE Spring Meeting, San Antonio, TX

13TH GLOBAL CONGRESS ON PROCESS SAFETY

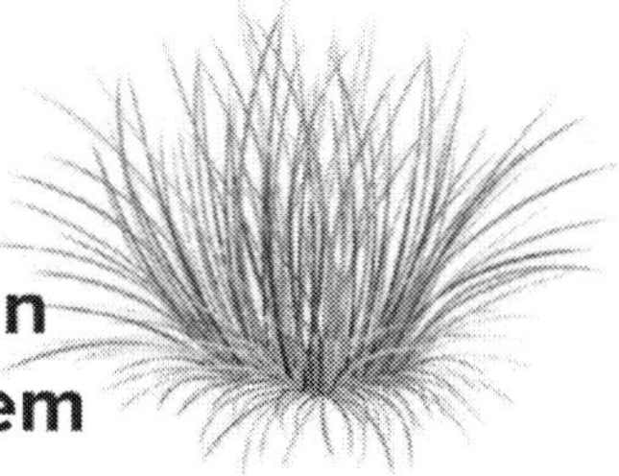

Carbon dioxide recapture utilizing an integrated vertical aquaponics system

Brittany Foley, Andrew Horowitz
Joe Laviano, Katy Wardzala
Northeastern University
Boston, MA

Increasing food demand exists worldwide

- Global food production will require a **50-60% increase** over the next 30 years (*LA Times*)
- It is estimated that ⅔ **of the world's population** will live in cities by 2050 (*"Growing Food for Growing Cities"*)
- **50-70% of the U.S.'s food supply** is consumed by people living in **urban communities** (*LA Times*)
- Food supply for **urban areas** is largely **unsustainable**
- Food today is often **shipped thousands of miles**

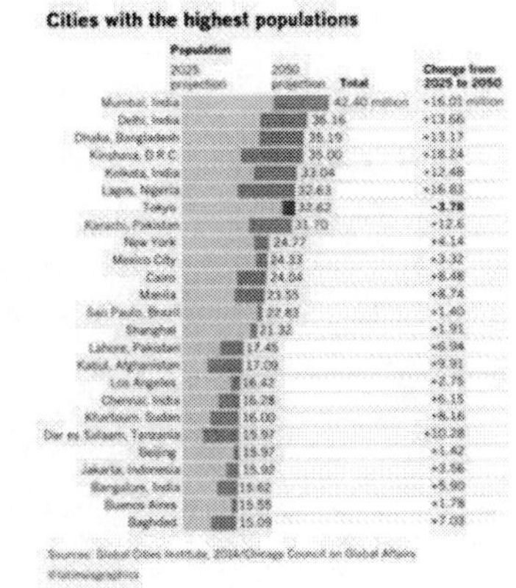

"The threat is, will there be enough food produced for those people living in cities?"

Dan Glickman, former U.S. Secretary of Agriculture

17AIChE Spring Meeting, San Antonio, TX | 13TH GLOBAL CONGRESS ON PROCESS SAFETY

Hydroponics

CityFARM. MIT, n.d. Web.

- No need for **pesticides, fungicides**
- **Year-round** crop cultivation
- Not weather-dependent
- Drastically reduces agricultural waste
- **Any unused space** can be turned into a vertical hydroponic farm
- *Challenge*: systems require fertilizer to be added on a daily basis

17AIChE Spring Meeting, San Antonio, TX | 13TH GLOBAL CONGRESS ON PROCESS SAFETY

Aquaponics

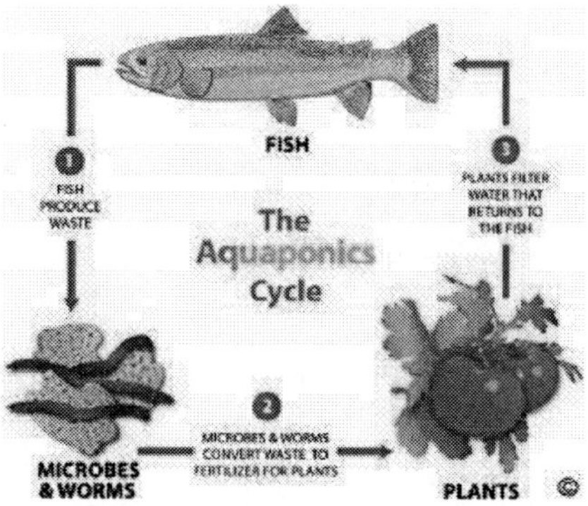

The Aquaponics Source. (n.d.). https://www.theaquaponicsource.com/what-is-aquaponics/

- A form of **hydroponics** that utilizes **waste** produced by fish to fertilize plant growth
- Farming of both **fish** and **plants**
- Fish may be ornamental (koi, etc.) or grown for food (i.e. tilapia, trout, etc.)
- **Fully continuous system**
- **Two inputs:** light and nutrients for fish

17 AIChE Spring Meeting, San Antonio, TX — 13TH GLOBAL CONGRESS ON PROCESS SAFETY

"In the last 20 to 30 years, the amount of food being transported from rural to urban areas has increased 300% in India, 800% in Africa and 1,000% in Southeast Asia."

Thomas A. Reardon, Chicago Council on Global Affairs, for *L.A. Times*

17 AIChE Spring Meeting, San Antonio, TX — 13TH GLOBAL CONGRESS ON PROCESS SAFETY

Carbon dioxide recapture utilizing an integrated vertical aquaponics system

Our system: A commercial vertical aquaponics system with integrated controls, using carbon dioxide produced from fermentation and heat recapture from LED grow lights.

Recycled Carbon Dioxide

- Enhancing growth with CO_2 but with reduced energy consumption relative to competitors who burn propane for CO_2
- Supply can be sourced from other systems' waste (i.e. breweries)

Heat Recapture

- Heat released by LED growth lights used to control water tank temperature at 75° ± 3°F

17 AIChE Spring Meeting, San Antonio, TX — 13TH GLOBAL CONGRESS ON PROCESS SAFETY

Demonstrative model

Fish tank for tilapia

Sump tank to maintain water level

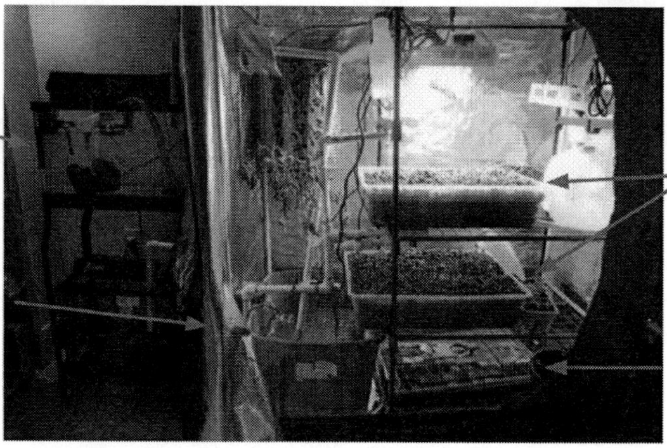

Two growth beds with expanded clay media

Seed starters in rockwool and humidity dome

17 AIChE Spring Meeting, San Antonio, TX — 13TH GLOBAL CONGRESS ON PROCESS SAFETY

Setting & Market

Target Market

- Small-scale breweries
- Alleviate waste CO_2 expenses
  - Provide incentives
  - Offer alternative income
  - Market popularity = expansive growth

Plant Location

- Greater Boston Area

Design of a 1,000 ft² scalable model

Model Assumptions

- 8,000 barrel/year production capacity
- Assumes beer brewed at 14 → 2 degrees plato
- Assumes ⅔ CO_2 recapture

Product Basis

- Cultivating 50:50 lettuce and basil
- Crop densities planted at 4/ft² - 12/ft²
- Short life cycles (8 weeks)
- Plant consumes ½ of its weight in CO_2 during its lifetime

Specifications

- Four quad-tier growth beds (125 ft²/bed)
- Four 1,200 gallon fish tanks (50 ft²/tank)
- 1000 ft² total occupied floor space

Major equipment specifications

- Media Beds
  - Food-grade containers
- Fish & Sump Tank
  - 1,200 gallon receptacles
- Heat Exchanger
  - Shell-and-tube design
- Pumps
  - Centrifugal
  - 1000 gph rated (assumes 75% packing)

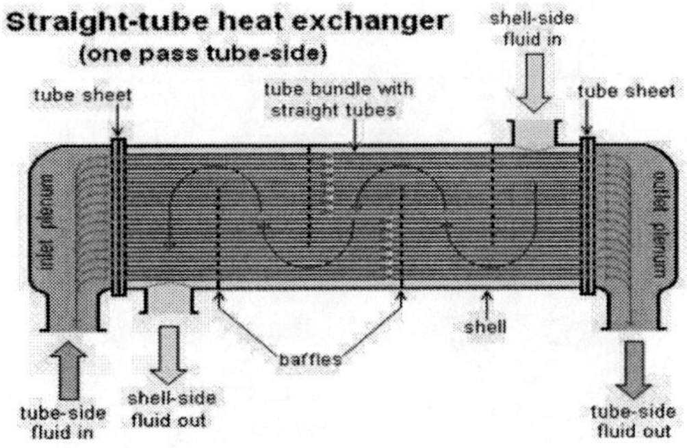

Control philosophy

Automated Control

- Pumps and Valves
- Temperature Control
- Lighting

Manual Control

- Seeding
- Sampling
- Harvesting
- Water Quality Control

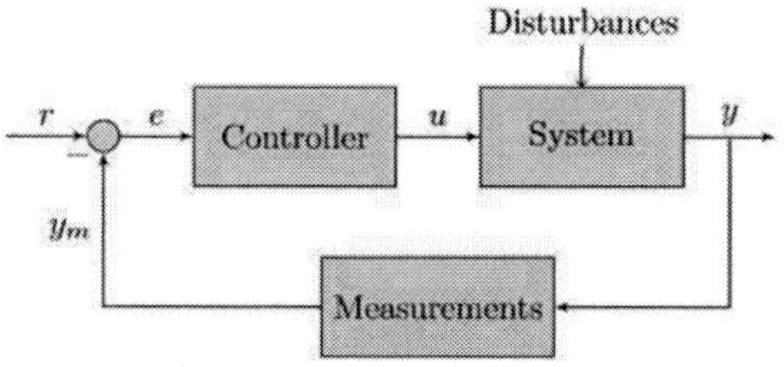

Automated controls

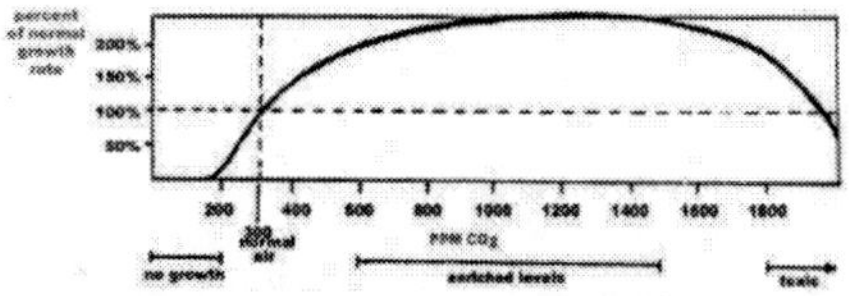

Temperature

- Logged once per second
- Alarms outside of 72-78°F
- Distributed Control System (DCS)

Carbon Dioxide

- Gas analyzer
- High alarm at 2,500 ppm
- Low alarm at 1,200 ppm

Flow

- Valves and pumps monitored by DCS

Level

Manual controls

Water quality will be continuously monitored, only requiring operator intervention when necessary.

Harvesting/Planting

- Plants: 4-8 weeks
- Tilapia: 10 months
- Staggered life cycles throughout

Daily Sampling

- Plant growth
- Fish health

Overall process flow

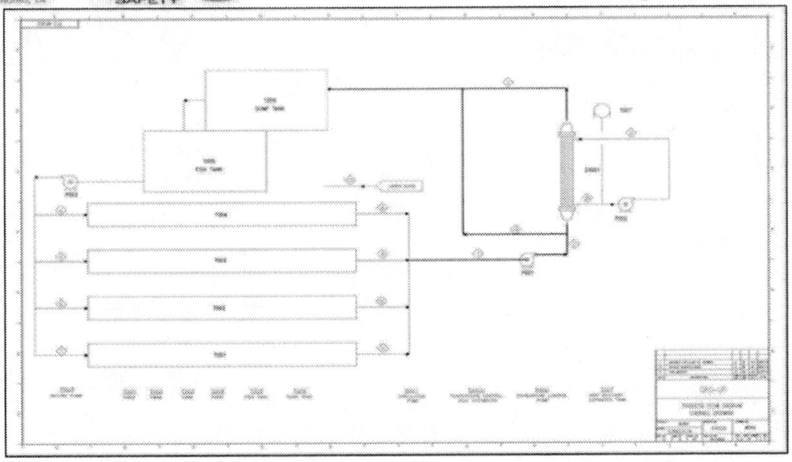

17AIChE Spring Meeting, San Antonio, TX — 13TH GLOBAL CONGRESS ON PROCESS SAFETY

Installed capital costs

(for client)

	Units	Units/Unit product	Quantity	Cost ($)/unit	Cost ($)
High-powered 300W LED grow lights	ea	1	16	109.00	1,744.00
Fish filter pump	ea	1	4	110.00	440.00
HDPE tubing 2"	ft	300	2	573.00	1,146.00
Metal supports (shelving)	ea	1	4	510.00	2,040.00
Water pumps for circulation system	ea	1	8	71.05	568.40
Air pump for oxygenation	ea	1	4	94.24	376.96
Air hoses	ft	1	80	2.00	160.00
Ammonium chloride	g	500	1	14.06	14.06
Fish and sump tanks	ea	1	8	724.00	5,792.00
Temperature transmitters	ea	1	16	30.35	485.60
Flow meter	ea	1	8	79.99	639.92
Composition analyzers	ea	1	12	185.00	2,220.00
Automated on/off valves	ea	1	36	46.65	1,679.40
Control valves	ea	1	12	53.42	641.04
Food-grade plastic containers for grow beds	ea	1	16	625.00	10,000.00
CO_2 Analyzer Thermo 41H	ea	1	1	600.00	600.00
TAC I/A Series: MNB-300 BACnet	ea	1	1	373.00	373.00
TAC I/A Series: MNB-V1-2	ea	1	1	216.30	216.30
TAC I/A Series: MNB-V2-2	ea	1	1	231.70	231.70
TAC I/A Series: MNB-1000	ea	1	1	1,080.45	1,080.45
FoxView & FoxDraw	ea	1	1	0.0	0.00
Computer	ea	1	2	499.0	998.00
(Table 3 cont.)	Units	Units/Unit product	Quantity	Cost ($)/unit	Cost ($)
Heat exchanger	ea	1	1	834.0	834.00
Air sealed tent, 1000 sq. ft.	ea	1	1	550.0	550.00
Total ISBL Costs					$32,830.83

- Economic model is based on brewery with an **8,000 barrel/year** production rate
- **Four** total cultivation units per system
- Number of units depends on:
    - Brewery size
    - Desired CO_2 recapture rate

17AIChE Spring Meeting, San Antonio, TX — 13TH GLOBAL CONGRESS ON PROCESS SAFETY

Fixed cost of production

(for client)

- 2 **Full-time operators** required
- Additional seasonal help as necessary (Internship opportunities)
- Size of project allows management to be absorbed by parent company

	Units	Units/year	Cost	Cost/year
Fish Care	kg	1,000.00	$0.50	$500.00
Clay media	ton	72.70	$60.00	$4,362.00
Seeds	2ea	104,000.00	$0.01	$1,308.32
Fish	ea	1,050.00*	$0.99	$1,039.50
Total Raw Materials (RM)				$7,209.82

			Cost/year
Labor			
	2.0	Operators per Shift Position	
# Shift positions	1.0	salaried at $36,800 ea.	$73,600.00
Maintenance			
		3% of ISBL Investment	$984.92
		GroUp Yearly Maintenance	$2,000.0
Overhead Expense			
Plant Overhead		40% of Labor & Maintenance	$29,833.97
Tax & Insurance		2% of Fixed Investment	$3,324.32
		Fixed Cost of Production (FCOP)	**$109,743.22**

17 AIChE Spring Meeting, San Antonio, TX — 13TH GLOBAL CONGRESS ON PROCESS SAFETY

Revenue streams

(for client)

- **Market value** ranges depending on consumer
  - $1.25 - $5.00 per unit
- **Three primary sources of distribution**
  - Wholesale distributors
  - Direct-to-consumer
  - Local restaurants

Key Products	Units	Units/unit product	Units /year	Cost/unit	Cost/year
Lettuce (Wholesale)	26,000	1	13,000	$1.25	$16,250.00
Lettuce (Direct)		1	6,500	$3.00	$19,500.00
Lettuce (Restaurants)		1	6,500	$2.50	$16,250.00
Basil (Wholesale)	78,000	2	78,000	$2.50	$195,000.00
Basil (Direct)		2	39,000	$5.00	$195,000.00
Basil (Restaurants)		2	39,000	$3.00	$117,000.00
Fish (Wholesale)		1	1,000	$3.00	$3,000.00
Total Key Product Revenues (REV)					**$562,000.00**

17 AIChE Spring Meeting, San Antonio, TX — 13TH GLOBAL CONGRESS ON PROCESS SAFETY

Projected growth

(for client)

Total Investment sums to $165,672

- Simple payback period of **0.57 years**
- 224.83% ROI after 10 years
- Engineering costs include installation, maintenance, training, and consulting fees
- Areas for sensitivity
  - Plant type
  - Price points

	Cost
ISBL Capital Cost	$32,482.03
OSBL Capital Cost	$6,496.41
Engineering Costs	$118,897.84
Contingency	$7,795.69
Total Fixed Capital Cost	**$165,671.97**
Working Capital	$16,567.20

17AIChE Spring Meeting, San Antonio, TX — 13TH GLOBAL CONGRESS ON PROCESS SAFETY

Economic incentives

(for client)

Direct Incentives:

- Tax Credits
  - Fish and Agriculture Income Tax Credit (Mass DOR) - 3% of production cost, including property
  - Lighting controls (MassSave) - \$0.50/ft^2
- Emissions Breaks - CO_2 recapture

Indirect Value Added:

- **Increased market value** for products grown hydroponically, locally, and pesticide free
- **Tourism/Novelty** of system

17AIChE Spring Meeting, San Antonio, TX — 13TH GLOBAL CONGRESS ON PROCESS SAFETY

Revenue breakdown

(GroUp)

Key Products/Services	Units	Units/yr	Price \$/unit	\$/yr
Plant Scale-Up/Scale-Down Spec	free	15	-	\$0.00
Scale Quote	free	8	0	\$0.00
Installation	fixed	4	\$100,000.00	\$400,000.00
Annual Service/Maintenance Charges	\$/yr	4	\$2,000.00	\$8,000.00
Training Services	fixed	4	\$10,000.00	\$40,000.00
Crop and aquaculture consultation	fixed	4*	\$2,500.00	\$10,000.00
Design modifications	fixed	1**	\$5,000.00	\$5,000.00
Additional maintenance charges	fixed	2**	\$5,000.00	\$10,000.00
Total Key Product Revenues		42.0		**\$473,000.00**

*Assuming all opt
**Estimated occurrence in a one-year period

17AIChE Spring Meeting, San Antonio, TX | 13TH GLOBAL CONGRESS ON PROCESS SAFETY

Business model
(GroUp)

Location: Charlestown, MA

Team: Four full-time chemical engineers employed at salaries of $65,000 annually

CoP Costs	$/yr
Variable Cost of Production	$33,926.40
Fixed Cost of Production	$365,801.86
Cash Cost of Production	$399,728.26
Gross Profit	**$73,271.74**

GroUp Costs	$
ISBL Capital Cost	$25,450.00
OSBL Capital Cost	$5,090.00
Engineering Costs	$3,054.00
Contingency	$3,054.00
Total Fixed Capital Cost	$36,648.00
Working Capital	$3,664.80

17AIChE Spring Meeting, San Antonio, TX | 13TH GLOBAL CONGRESS ON PROCESS SAFETY

Acknowledgements

We would like to offer thanks to...

AIChE & the 2017 Spring Meeting team for welcoming us to this year's meeting in San Antonio and inviting us to present.

Our advisors Thomas Webster, Devin Hersey, Don Wood, and Amit Roy for their leadership and guidance in our Capstone course.

Northeastern University and the Northeastern Chemical Engineering Department for making it possible for us to present today.

Nightshift Brewing Company, Downeast Cider House, and Harpoon Brewery for their interest, information, and support.

17 AIChE Spring Meeting, San Antonio, TX

13TH GLOBAL CONGRESS ON PROCESS SAFETY

Questions?

479802 Economic Optimization and Hazard Analysis of a Fluidized Catalytic Cracker

Abstract:

A fluidized catalytic cracking process was modeled along with regenerator and subsequent fractionator. The system was modeled using the process simulator ChemCAD and priced using cost equations from Turton, Bailie, Whiting and Shaeiwitz (2012). Decisions of representative feed and product compositions along with reaction pathways were made based on literature and requirement of model fidelity for economic analysis. Kinetics were regressed from industrially representative literature data. Both topological and parametric optimizations were considered. Process hazards will be identified and suggestions to mitigate hazards will be presented.

480893 Miniaturized Biomass Gasification System for Onboard Applications

Abstract:

Existing biomass gasification systems used for transportation are bulky, slow to start, and not much more efficient than traditional gasoline engines. Grassoline utilizes a small, efficient, and rapid-heating process to generate carbon-neutral electrical power for onboard applications. The design includes a packed-bed gasification reactor, a solid oxide fuel cell, and systems for fuel handling and electrical processing. The performance of a 1:18 scale model of the system is simulated in Aspen Plus.

2017 AIChE®

SPRING MEETING&

13TH

GLOBAL CONGRESS

ON PROCESS SAFETY

March 26–30, 2017
Henry B. Gonzalez
Convention Center
San Antonio, TX

17AIChE
Spring Meeting, San Antonio, TX

13TH GLOBAL CONGRESS ON PROCESS SAFETY

17AIChE
Spring Meeting, San Antonio, TX

13TH GLOBAL CONGRESS ON PROCESS SAFETY

Value of Modern Automation in Process Industries

Laurie R. Ben
Director, Global Business Development
Emerson Automation Solutions

17AIChE Spring Meeting, San Antonio, TX — 13TH GLOBAL CONGRESS ON PROCESS SAFETY

Agenda

- Setting the Stage – Fundamentals of automation
- Why Modernize?
- Where are the Benefits? Follow the Money!
- Next Steps
- Real World Example
- Summary

17AIChE Spring Meeting, San Antonio, TX — 13TH GLOBAL CONGRESS ON PROCESS SAFETY

Automation Definition

- Consider Automation as being made up of:
  - Measurement Devices
  - A process control system to run control strategy
  - Final control elements

2017 AIChE® SPRING MEETING & 13TH GLOBAL CONGRESS ON PROCESS SAFETY

17 AIChE Spring Meeting, San Antonio, TX — 13TH GLOBAL CONGRESS ON PROCESS SAFETY

Fundamental Beliefs Why Automation Provides Value

- Automation is used to make process equipment operate at expected performance
- Most process units are constructed with a less than optimal amount of measurement & control
- Installed automation technology is not currently used to full potential
- Minor process unit equipment or piping modifications may be required along with automation to improve operational performance

17 AIChE Spring Meeting, San Antonio, TX — 13TH GLOBAL CONGRESS ON PROCESS SAFETY

Strategic Importance of Process Automation

- Increasing complexity and fewer resources increases reliance on technology
- Automation is only 3-5% of original CAPEX, but a huge factor in operational success

- Process automation touches every element of the process
- Successful implementation of process automation is key to the ongoing success of the operation

2017 AIChE® SPRING MEETING & 13TH GLOBAL CONGRESS ON PROCESS SAFETY

Where Benefits Come From

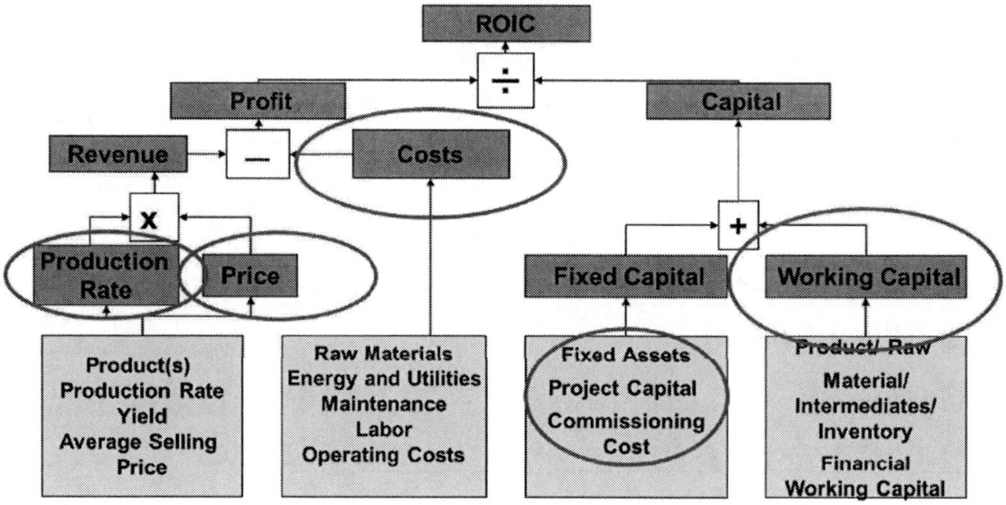

Why Modernize? Where are you on the curve?

Two types of costs associated with delaying modernization projects:

- Increasing Legacy System Health Costs
  - Parts from vendor, 3rd party, sister facilities, ebay becoming difficult, expensive, DOA,
  - Legacy expertise support from vendor, consultants, internally becoming limited
  - Site skills (operator, engineer, I&E) dwindling
- Decreasing Legacy System Automation Value (lack of competitive tools & technologies to meet business challenges)
  - Flexibility to address operational impacts/needs
  - Ability to address regulations, competitive pressures, cybersecurity, etc.
  - Expansion of process or system equipment
  - Expand employee capabilities & effectiveness

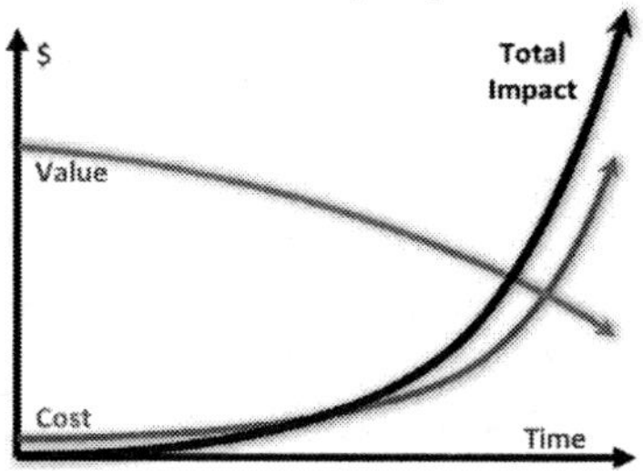

"It is not enough to just reduce the number of failures of obsolete hardware components."

"The justification must provide both cost and expected benefits (value) relative to the next best investment option."

ARC Advisory Group

17AIChE Spring Meeting, San Antonio, TX | 13TH GLOBAL CONGRESS ON PROCESS SAFETY

State of the Art System Benefits

- Sustainability
  - Mature System Support Costs
  - Long Term Maintainability
- Plant & Process Unit Improvement Opportunities
- Improved Production Management
- Improved Capital Efficiency & Maintenance Spending
  - Asset Management
  - Process Automation System
- Improved Safety & Environment Incidents
- Overall Equipment Effectiveness / Improved Reliability
- Improved Logistics & Supply Chain Management

17AIChE Spring Meeting, San Antonio, TX | 13TH GLOBAL CONGRESS ON PROCESS SAFETY

Sustaining Issues & Opportunity

- Concern of unscheduled downtime / incident do to obsolescence
- No longer cost effective to support old system
  - 30% of spare parts salvaged form older systems fail*
- Legacy system can not support new business opportunities
- Lack of capacity to expand
- Regulatory requirements
- Loss of operational intellectual talent through retirement

Legacy System MCC I/O

Modern MCC I/O

Plant & Process Unit Opportunities

Process Improvements

- Increased Capacity
- Production Quality
- Process Optimization

Poor Performance Causes

Variability

- Poor Control (Estimates of 2/3 of control loops under perform)
- Operators run with "safety" factory from constraint or specification
- Inconsistent batch operations
- Insufficient information for operating decisions

Deteriorating Conditions

- Equipment surface coating
- Catalyst degradation
- Equipment wear

Manual Operations

Plant Availability

Increased Capacity Batch Processes

Possible Benefits

- Largest Financial Gain if Sold Out
- Excess capacity to handle large demand periods
- Emergency orders
- Alignment of production assets between made to order and made to schedule product lines

Automation Activities to Obtain

- Lower Batch Cycle times
- Ability to Effectively Monitor based on measurements vs time
- Availability of Process Equipment
  - Shared
  - In maintenance
  - Eliminate/reduce 'waiting for operator initiated actions'
- Improved scheduling of raw materials and available tankage

Batch Production Quality 'Right the First Time'

17AIChE Spring Meeting, San Antonio, TX — 13TH GLOBAL CONGRESS ON PROCESS SAFETY

Possible Benefits

- Eliminates batch adjustments, major source of inconsistent cycle times
- Improved raw material yield
- Lower usage of 'processing aid' chemicals
- Reduced energy consumption
- Reduce utility usage (cooling water, chilled water / refrigerants, nitrogen)
- Frees up equipment for more production capacity
- Frees up tankage and eliminates labor & energy for rework

Automation Activities to Obtain

- Accurate raw material addition
- Improved batch reaction control, may vary
- Limit operator manual actions to reduce human error
- Electronic measurement of small manual additions with mass totals at specific time stamps

Improved Safety & Environment Incidents

17AIChE Spring Meeting, San Antonio, TX — 13TH GLOBAL CONGRESS ON PROCESS SAFETY

Possible Benefits

- Keeping people safe
- Reduced accidents and fines
- Reduced reporting requirements and investigations
- Increased equipment availability
- Improved good will within the public community
- Reduced equipment repair cost by early notification of equipment failure

Automation Activities to Obtain

- Correct measurements of non-process control item
- Proper process interlocking including conditional interlocking
- Use of batch hold and abort functionality
- Proper design and application of alarm strategy

2017 AIChE® SPRING MEETING & 13TH GLOBAL CONGRESS ON PROCESS SAFETY

17AIChE Spring Meeting, San Antonio, TX

13TH GLOBAL CONGRESS ON PROCESS SAFETY

Automation Project Process

APPRAISE Evaluate Options	SELECT Solution Options	DEFINE Develop Solution	EXECUTE & Evaluate	OPERATE & Maintain
• FEL1/Feasibility	• FEL2/Conceptual Engineering/Pre-FEED	• FEL3/FEED	• Detailed Engineering	

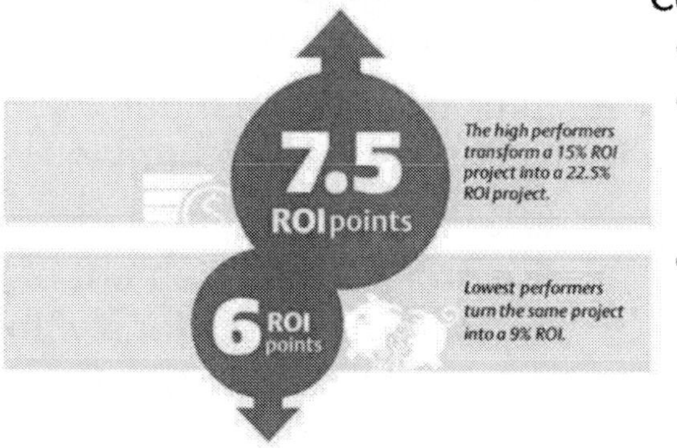

Source: Independent Project Analysis (IPA) Organization
Basis: $50M project with industry average ROI 15%

Common causes of decline

- Poor control in early planning
- Inadequate match between project and business objectives
- Paying too much for capital projects without entirely achieving business requirements

17AIChE Spring Meeting, San Antonio, TX

13TH GLOBAL CONGRESS ON PROCESS SAFETY

Design for the Justification Driver

- Determine what operational issues affect the financial KPIs
- Audit existing process for poor performance and variability
- Define opportunities for improvement through automation
- Determine what percent of the performance gap can be improved
- Estimate where the process should be running against these measures

2017 AIChE® SPRING MEETING & 13TH GLOBAL CONGRESS ON PROCESS SAFETY

Justification Identification

17AIChE Spring Meeting, San Antonio, TX — 13TH GLOBAL CONGRESS ON PROCESS SAFETY

- Need to work with:
  - Plant Management
  - Operations
  - Planning / Scheduling / Traffic
  - Marketing and Sales
- Need to work with Accounting group to understand plant cost accounting
  - Cost allocations per unit produced
  - Activity based cost accounting
  - Sales price
  - Production costs

Pre-FEED Example: FEL 2 Leads To Successful Project

17AIChE Spring Meeting, San Antonio, TX — 13TH GLOBAL CONGRESS ON PROCESS SAFETY

- CHALLENGES
  - High batch failure rate: 18 / 600
  - Low capacity utilization
  - Materials consumption high
  - First pass quality below targets
- STRATEGY (3-Phased Approach)
  1. Digital instrumentation & control
  2. Model Predictive Control
  3. Batch Automation
- RESULTS
  - 800+ consecutive batches
  - 33% increase in throughput
  - 7% increase in 1st pass quality
  - 56% less time spent reviewing Quality Assurance documentation
  - 80% reduction in product release documentation
  - Plans for expanded production line on hold, no need to build new process equipment units!

EMERSON

2017 AIChE® SPRING MEETING & 13TH GLOBAL CONGRESS ON PROCESS SAFETY

17AIChE Spring Meeting, San Antonio, TX — 13TH GLOBAL CONGRESS ON PROCESS SAFETY

Pre-FEED Assessment Results - Examples

Benefits identified through automation assessment in Pre-FEED

Modernization Master Plan Is Used To Justify, Plan & Prioritize Projects

- $8 M annualized modernization benefits
- >$5M TICC estimate for 15 units
- Clarified operational procedures, issues, & future product demand
- Quantified batch cycle time variability by studying batch data
- Itemized instrumentation and controls equipment through site audit
- Prescribed an automation strategy to reduce batch cycle time variability

Modernization Opportunity Assessment Service Helps Prioritize Projects

- $30M TICC estimate: $1M budget recommended for FEED
- Advanced controls to save $500k in energy costs
- Continuous process sequential automation to improve capacity, saves >$100k energy costs & eliminates >$200k annual repair costs due to operator error
- Batch operational improvements range from $100k to over $1M

17AIChE Spring Meeting, San Antonio, TX — 13TH GLOBAL CONGRESS ON PROCESS SAFETY

Summary

- Fundamentals of Automation
- Why Modernize – Sustaining and ROI
- Example Benefits & Automation Activities
- Best Practice of Justification Process
- Questions

2017 AIChE® SPRING MEETING & 13TH GLOBAL CONGRESS ON PROCESS SAFETY

480835 Automatic Control of Polymer Molecular Weight during Free Radical Polymerization

Abstract:

A model-free, fully automatic feedback molecular weight controller was developed, which uses Automatic Continuous Online Monitoring Polymerizations (ACOMP) data. ACOMP provides continuous real-time monomer and polymer concentration, weight average molecular weight and intrinsic viscosity. The model-free controller uses two proportionality constants for monomer concentration and molecular weight control; i. alpha, monomer consumption rate constant and ii. p, monomer concentration proportional to weight average kinetic chain length. The controller has been successful for molecular weight control of linear chain growth free radical polymerizations. Using the ACOMP measurements, the instantaneous weight average molecular weight can be computed and folded into a chromatography-free molecular weight distribution (MWD). Hence, the evolution of the MWD can be monitored in real time and actively controlled. Multimodal distributions were achieved through the control of monomer and chain transfer agent concentration. Chain transfer agent is an optimal control for molecular weight because it affects chain length but not reaction kinetics. Furthermore, the APMT/PolyRMC controller has successfully controlled co-monomer composition and molecular weight by computing comonomer conversion kinetics and instantaneous comonomer fraction.

Automatic, Simultaneous Control of Polymer Composition and Molecular Weight during Copolymer Synthesis

Authors: Terry McAfee(1), Rick Montgomery(2), Wayne Reed(1), Thomas Zekoski(1), Aide Wu(1)

(1) Tulane University

(2) Advanced Polymer Monitoring Technologies

PolyRMC

The ACOMP/Control Interface

Towards full feedback control of polymerization reactions

APMT Pilot System: 64 L reactor w/ 37.85 lpm fast loop

PolyRMC System: 2L reactor w/15-20 ml/min fast loop

Implemented June 2015

2

Successful Control of Monomer Concentration

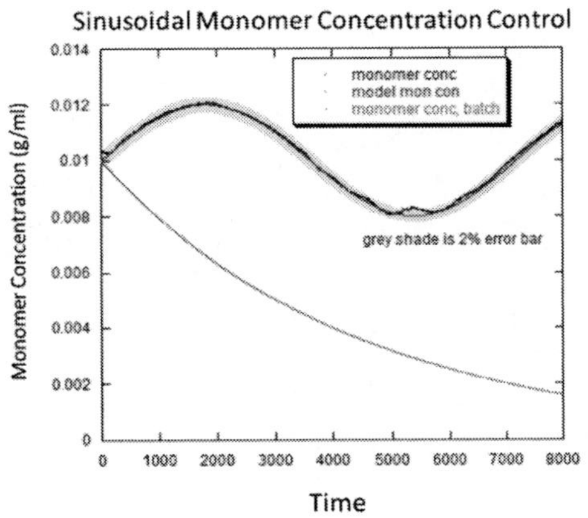

3

Successful Control of Molecular Weight

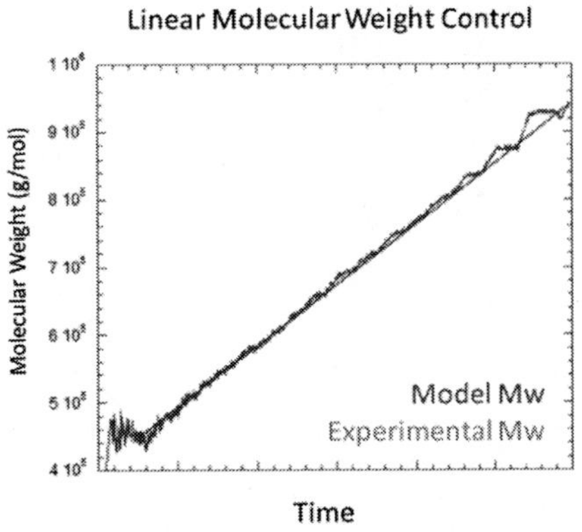

4

ACOMP's Real-time $M_{w,inst}$ holds insightful information otherwise unobtainable

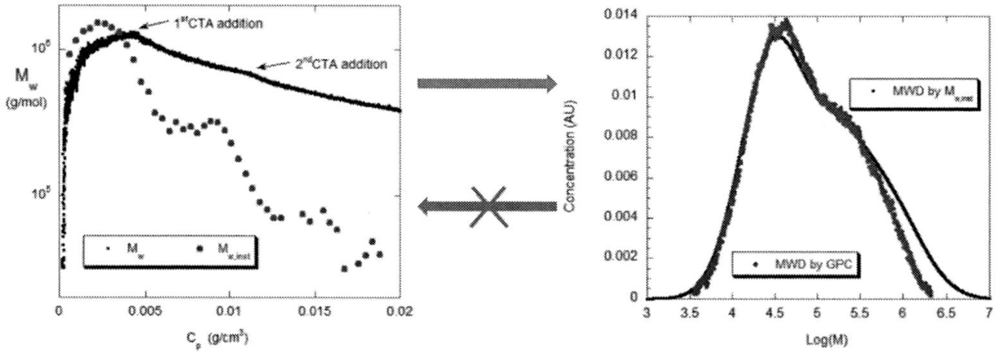

Terpolymerization: NMR spectra of separate monomers and combined comonomers: Styrene sulfonate, sodium Acrylate, Acrylamide

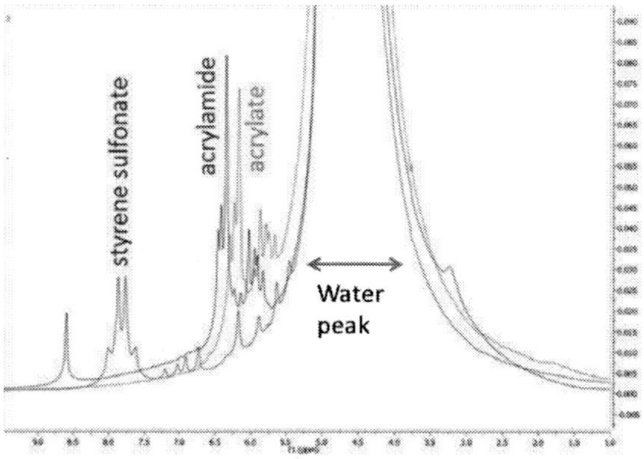

Terpolymerization data: Styrene sulfonate, sodium acrylate, acrylamide

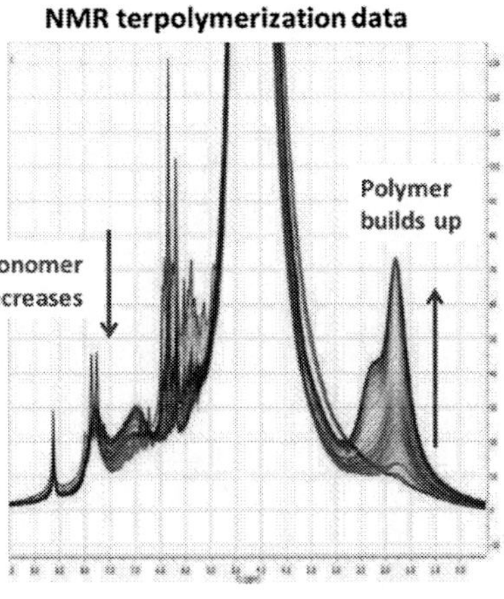

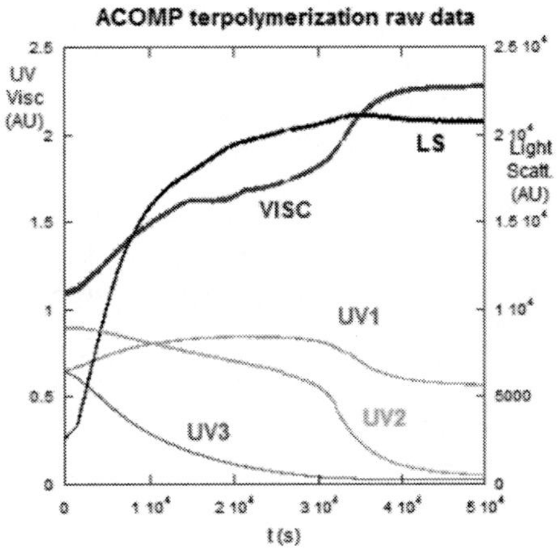

First Fully Automated Composition Controller for Copolymer Synthesis

- Does not require specific kinetic models, reactivity ratios, etc.
- Requires only knowledge of α_A and α_B, which can be directly measured throughout reactions

$$M_{w,inst} = pm \qquad m = A + B$$

$$F_A = \frac{\alpha_A A}{\alpha_A A + \alpha_B B} \qquad \Delta A_p = \alpha_A A \Delta t \qquad \Delta B_p = \alpha_B B \Delta t$$

Automatically set comonomer feed pump rates to

$$Q_A(t) = \frac{V(t)}{C_A{}'}\left(\frac{\Delta A}{\Delta t} + \alpha_A A\right)$$

$$Q_B(t) = \frac{V(t)}{C_B{}'}\left(\frac{\Delta B}{\Delta t} + \alpha_B B\right)$$

Copolymerization of Acrylamide and Styrene Sulfonate

F_1 = Instantaneous Fraction of SS
C_m = Total Monomer Concentration
C_p = Total Polymer Concentration

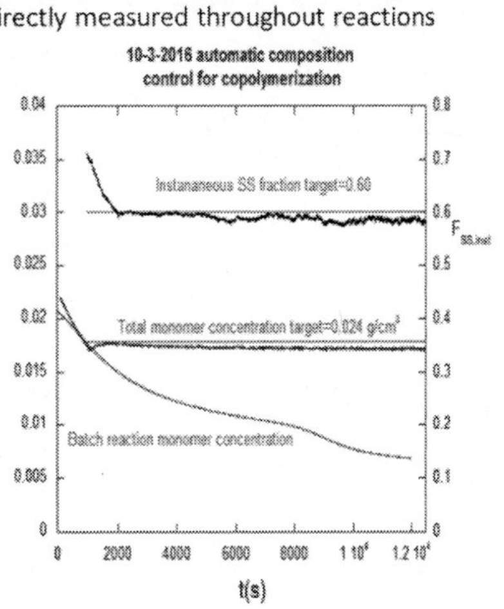

8

Simultaneous Mw and composition controller

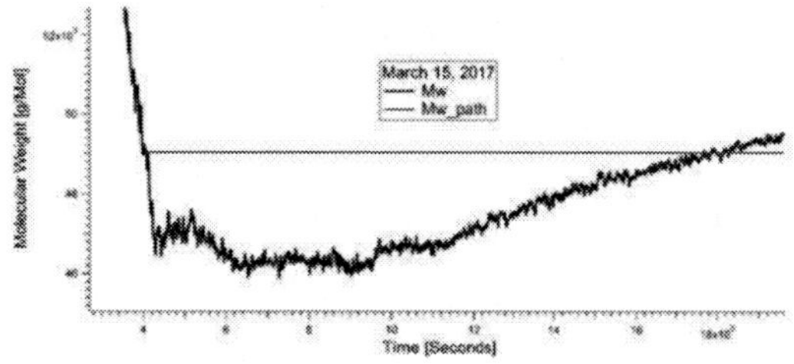

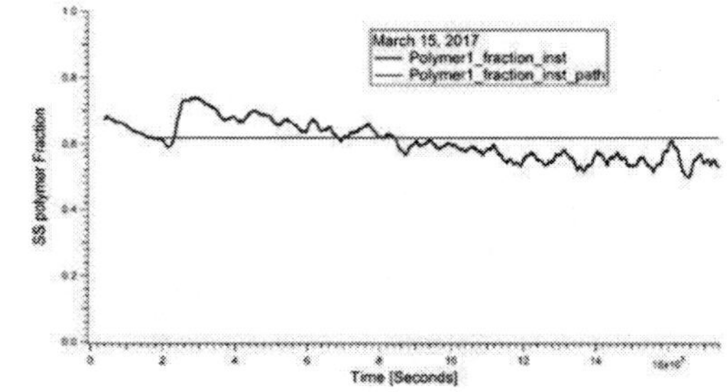

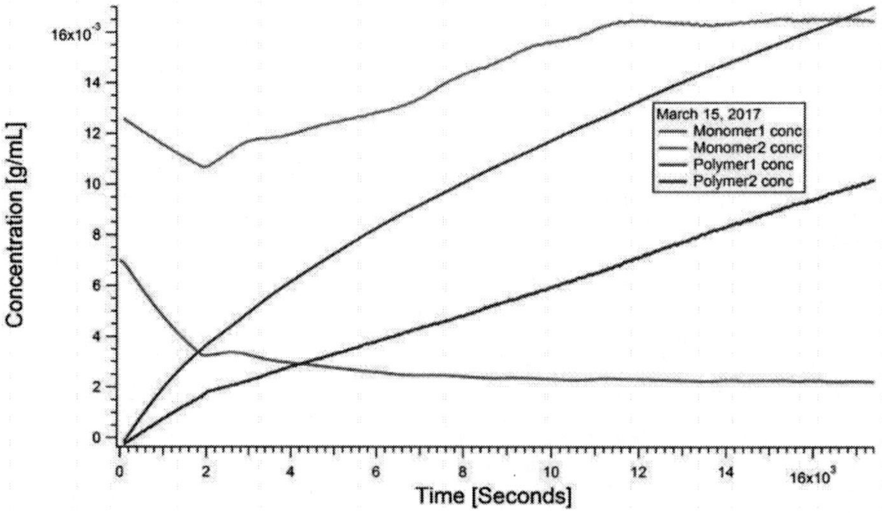

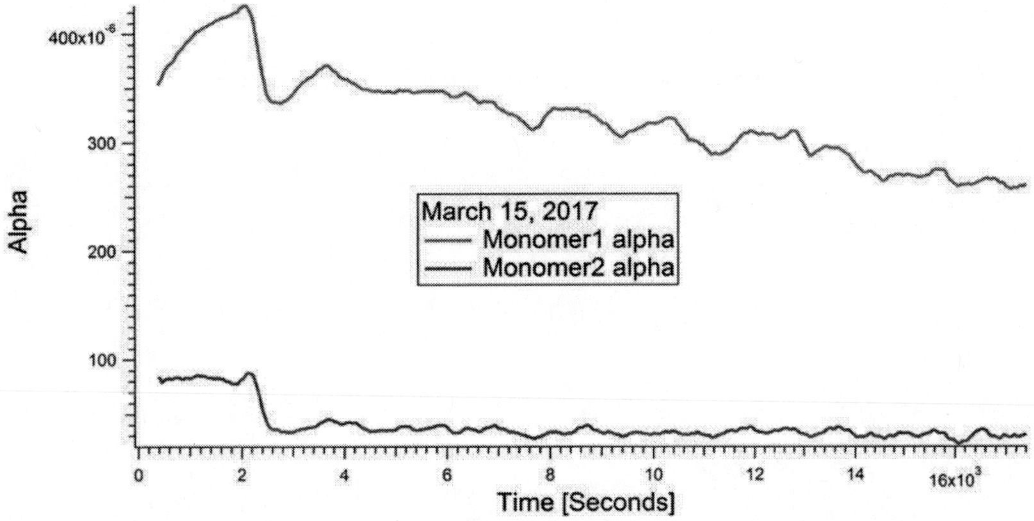
400x10⁻⁶
300
200
100
Alpha
March 15, 2017
Monomer1 alpha
Monomer2 alpha
0
2
4
6
8
10
12
14
16x10³
Time [Seconds]

Good Process Control Improves Plant Availability and Reliability

James Beall
Principal Control Consultant
Emerson Automation Solutions

Topics

- Control issues that can affect reliability
- Effective ways to deal with equipment limits
- Systems to monitor and warn operators when processes are outside desired operation window

17 AIChE Spring Meeting, San Antonio, TX — 13TH GLOBAL CONGRESS ON PROCESS SAFETY

How Does Poor Control Affect Equipment Reliability?

Examples:

- Pump cavitation
- Furnace coking
- Process and equipment variability
- Operating outside equipment design
- Increased shutdowns

17 AIChE Spring Meeting, San Antonio, TX — 13TH GLOBAL CONGRESS ON PROCESS SAFETY

Control Performance Improvements

- 4-8% increase in throughput
- 5-10% reduction in energy costs
- 2-8% reduction in product inventories
- 40-80% reduction in quality variation
- 1-5% increase in equipment availability

2017 AIChE® SPRING MEETING & 13TH GLOBAL CONGRESS ON PROCESS SAFETY

17 AIChE Spring Meeting, San Antonio, TX | 13TH GLOBAL CONGRESS ON PROCESS SAFETY

Pump Cavitation

- Common phenomenon with centrifugal pumps
- Caused by the formation and subsequent implosion of bubbles inside the pump casing.
- Results when Net Positive Suction Head (NPSH) is too low
- Related to the vapor pressure of the liquid at the suction conditions
- Causes physical damage to the impellers and pump casing.

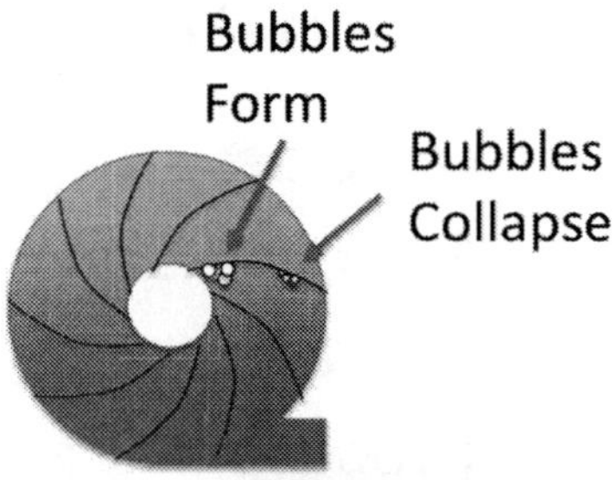

Source:
http://www.pumpfundamentals.com/

17 AIChE Spring Meeting, San Antonio, TX | 13TH GLOBAL CONGRESS ON PROCESS SAFETY

What Causes Cavitation?

- Suction drum, column or tank level too low
- Inlet temperature is high
- Plugging suction strainer
- Flow too high (e.g. high line pressure drop)
- Operating parallel pumps
- Any combination of the above

2017 AIChE® SPRING MEETING & 13TH GLOBAL CONGRESS ON PROCESS SAFETY

17 AIChE Spring Meeting, San Antonio, TX — 13TH GLOBAL CONGRESS ON PROCESS SAFETY

What Can Control Do About Cavitation?

- Stabilize level and temperature control
- Minimum level warnings
- Minimum pump suction pressure constraints
- Calculate and control minimum NPSH
- Flow balance multiple pumps
- Monitor for cavitation

17 AIChE Spring Meeting, San Antonio, TX — 13TH GLOBAL CONGRESS ON PROCESS SAFETY

Furnace Coking

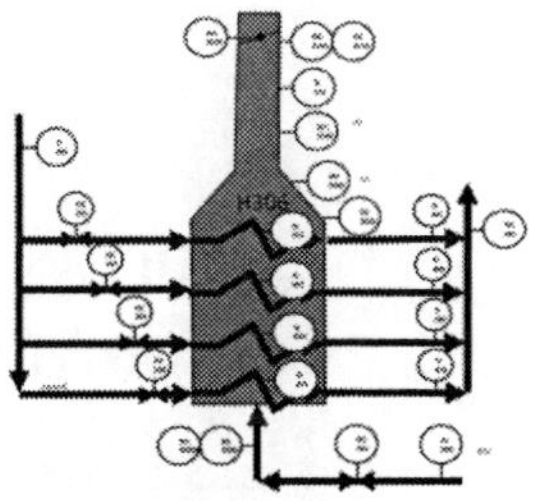

- Common issue for refinery furnaces
- Thermal cracking of hydrocarbon to coke which builds up on the inside of the tubes
- Reduces furnace efficiency
- Reduces heat transfer (i.e. lower outlet temp)
- Eventually requires tube decoking

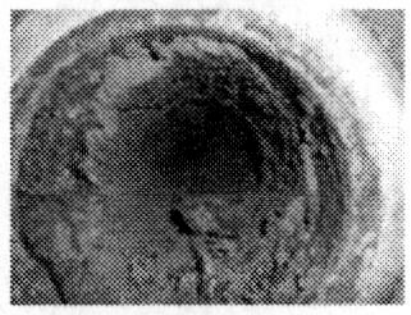

Source: http://www.asiaprotech.com/emainset.htm?id=pro03&xcode=03

17AIChE Spring Meeting, San Antonio, TX | 13TH GLOBAL CONGRESS ON PROCESS SAFETY

What is Coking Affected by?

- Tube temperatures
- Hot spots
- Hydrocarbon feed composition (longer chains crack easier)
- Residence time
- Steam injection
- Flame impingement

17AIChE Spring Meeting, San Antonio, TX | 13TH GLOBAL CONGRESS ON PROCESS SAFETY

What Can Control Do About Coking?

- Stability!!! Pass flows, fuel, steam, outlet and tube skin temperatures
- Pass balancing – Minimize pass temp differences
- Fuel BTU compensation – Use Coriolis for fuel flow measurement!
- Constraint control – Tube skin temp, pass temp differences, firebox, cross-over, tube pressure drop

11

17 AIChE Spring Meeting, San Antonio, TX — 13TH GLOBAL CONGRESS ON PROCESS SAFETY

Process Variability

- Control loop variability interacts throughout the process
- Oscillations affect downstream operations too
- Variability creates waste – Energy, quality, throughput, yield
- From reliability perspective, variability is hard on equipment

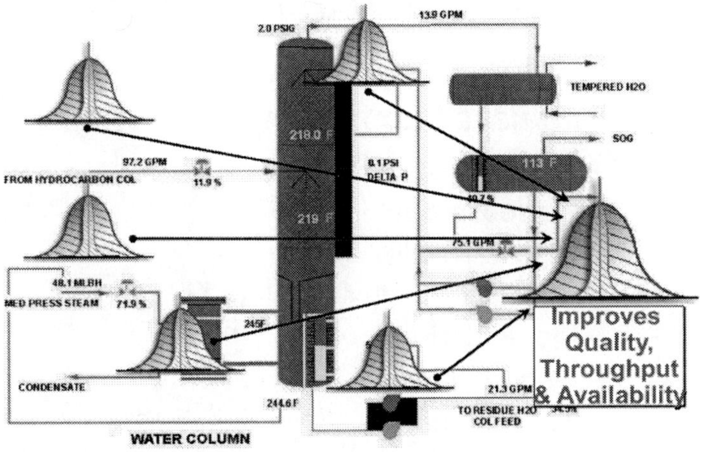

2017 AIChE® SPRING MEETING & 13TH GLOBAL CONGRESS ON PROCESS SAFETY

17 AIChE Spring Meeting, San Antonio, TX — 13TH GLOBAL CONGRESS ON PROCESS SAFETY

Variability Management

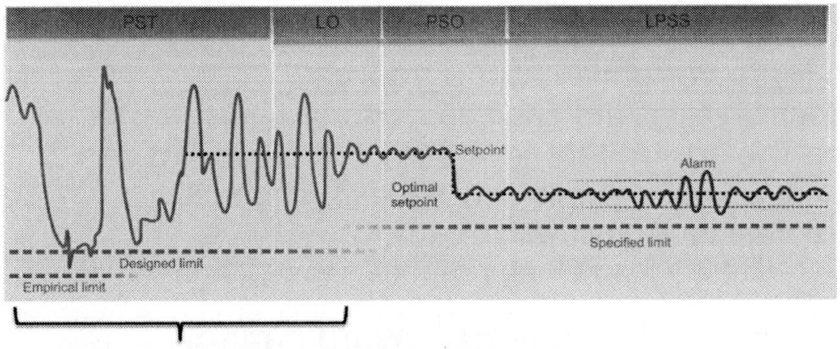

This kind of operation is hard on equipment!

2017 AIChE® SPRING MEETING & 13TH GLOBAL CONGRESS ON PROCESS SAFETY

17 AIChE Spring Meeting, San Antonio, TX — 13TH GLOBAL CONGRESS ON PROCESS SAFETY

Control Loop Monitoring Tools

- Embedded DCS tools
- Automatic Loop Tuning
- Control Loop KPI's
  - Correct mode
  - High variability
  - Oscillation index
  - Inputs bad
  - Output limited
  - Tuning index

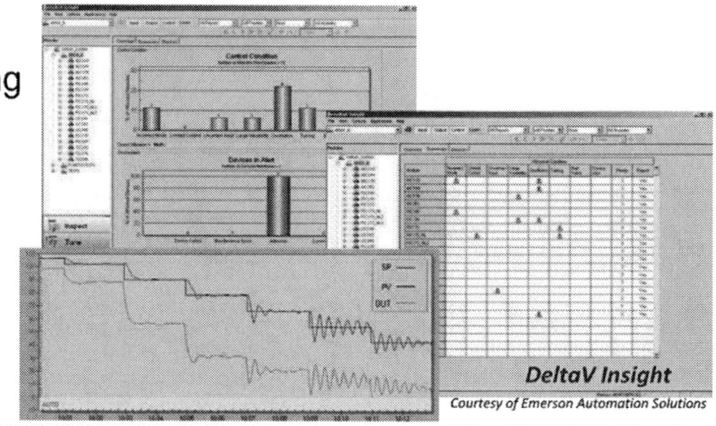

Courtesy of Emerson Automation Solutions

17 AIChE Spring Meeting, San Antonio, TX — 13TH GLOBAL CONGRESS ON PROCESS SAFETY

Control Performance Methodology

- Eliminate variability at the source
  - Control valves are often bad actors!
- Tune the controllers to meet control objectives
  - Most tuning methods only have "one speed"
  - Lambda tuning allows
  - Coordinate closed loop response to maximize PROCESS performance
- Utilize cascade and feed forward control
- Utilize model predictive, multivariable control (MPC)
- Minimize the affect of loop resonance

![17 AIChE Spring Meeting, San Antonio, TX — 13th Global Congress on Process Safety]

Asset Monitoring Tools

- Determine current health of equipment
- Monitor for excursions outside design operating window
- Statistically-based Analytics
- 1st Principals Models
- Notifications and alerts
- Dashboards
- Web-based implementations

Pump Failure Mode Analysis

Fault Condition	Vibration	Peak Impacting	Pressure (Discharge, Differential and Seal Fluid)			Seal Fluid Level	Temperature	Leak Detection
High Vibration	☑							
Cavitation		☑	☑					
Bearing fault		☑						
Pre-cavitation			☑					
Low head			☑	☑				
Low Discharge			☑					
Seal Pressure				☑				
Low Suction				☑				
Low Flow				☑				
Strainer Fault				☑				
Seal Failure					☑			
Seal Fluid Level						☑		
Bearing temp							☑	
Liquid HC Leak								☑

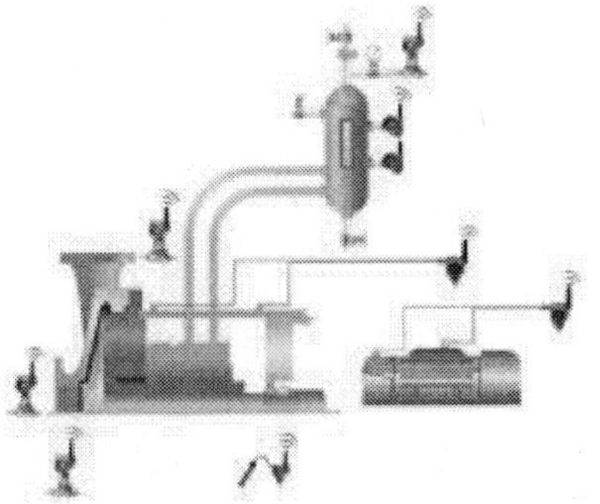

Equipment Health Indicator

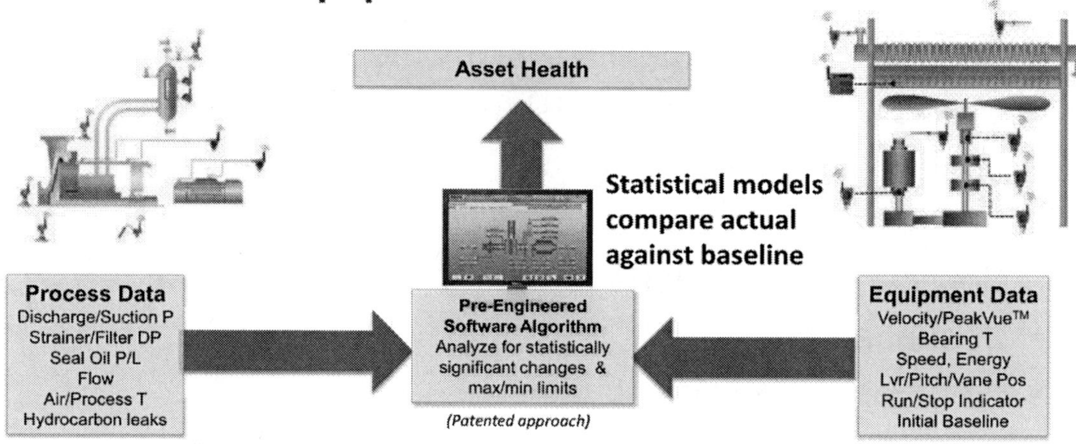

Example Health Monitoring Dashboard

- Plant Hierarchy navigation
- Scalable across 1000's of assets
- Easy identification of issues
- Easy drill down
- Automatic notification (to the right person) of active alerts

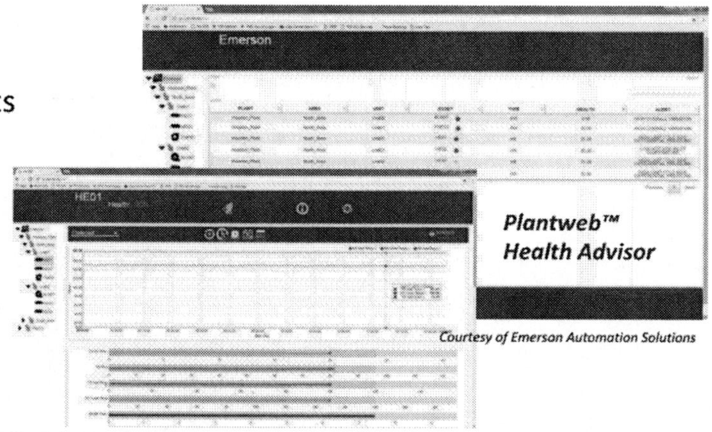

Courtesy of Emerson Automation Solutions

Health Monitoring Allows You To...

- Identify process problems causing equipment failures
- Receive early warning on equipment degradation
- Prevent leaks, fires and emissions
- Reduce repeat failures of equipment
- Reduce unplanned maintenance costs
- Perform only required maintenance
- Quickly identify abnormal situations

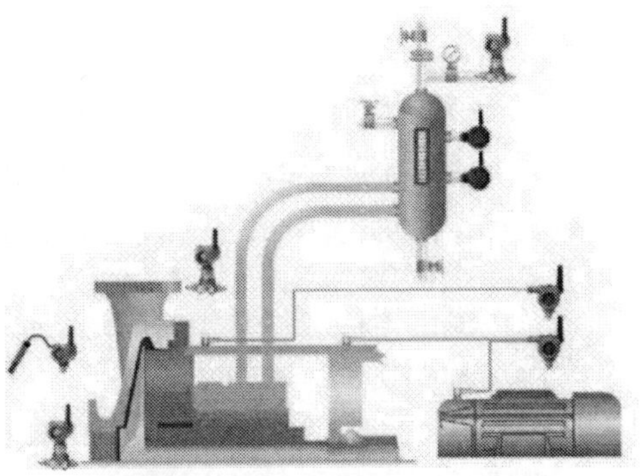

17 AIChE Spring Meeting, San Antonio, TX
13TH GLOBAL CONGRESS ON PROCESS SAFETY

Performance Monitoring Tools

- 1st Principal models to compare actual operation vs. design
- Key metric is efficiency & cost of degradation
- Requires equipment design curves & more measurements
- More expensive to implement
- Usually reserved for more critical equipment

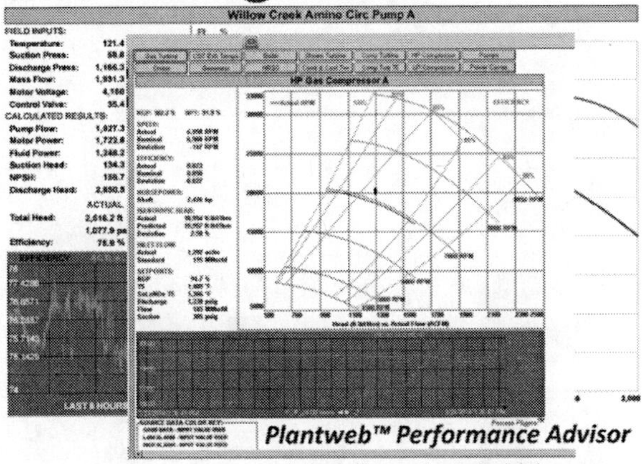

Courtesy of Emerson Automation Solutions

EMERSON

2017 AIChE® SPRING MEETING & 13TH GLOBAL CONGRESS ON PROCESS SAFETY

17 AIChE Spring Meeting, San Antonio, TX — 13TH GLOBAL CONGRESS ON PROCESS SAFETY

Performance Monitoring Allows you to…

- Get immediate notification if any monitored equipment has a fault affecting performance
- Understand degradation trends and status
- Identify potential instrument problems, pinpoint degradation sources, and evaluate the effectiveness of cost improvement actions
- **Understand financial impact of performance deviations**

17 AIChE Spring Meeting, San Antonio, TX — 13TH GLOBAL CONGRESS ON PROCESS SAFETY

The Future is in the Cloud

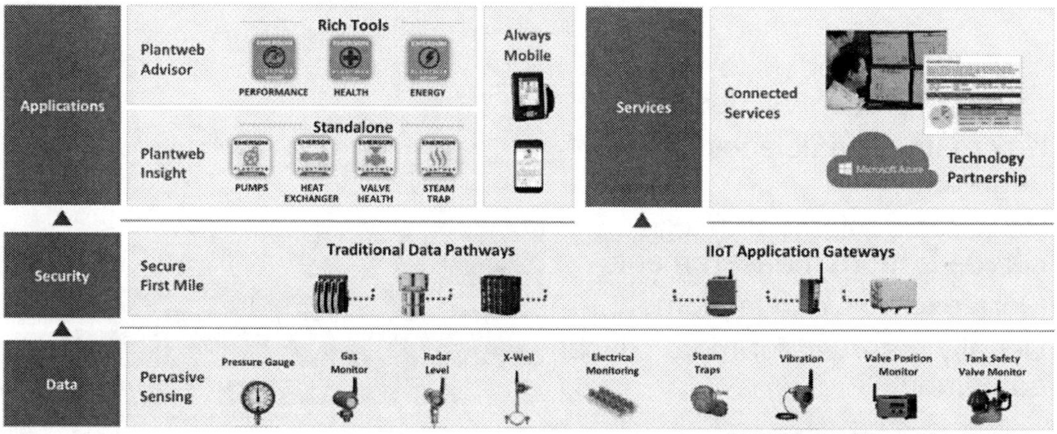

Reprinted courtesy of Emerson Automation Solutions

Summary

- Process control performance impacts equipment reliability/availability
  - Less variability on process and equipment
  - Less frequent shutdowns
- Asset monitoring systems
  - Identify process problems causing equipment failures
  - Receive early warning on equipment degradation
- 1-5% Improvement in availability/reliability

480336 Utilization of Flare Gas for Direct Non-Oxidative Production of Liquid Fuels and Chemicals

Abstract:

Natural gas flaring produces 400 million tons of greenhouse gas emissions annually [1]. The current gas-to-liquids technology such as Fischer-Tropsch process requires two stages with synthesis gas ($CO + H_2$) as intermediate, which is extremely capital-intensive and not economically viable for stranded gas utilization. Other technology, e.g., the oxidative coupling process is difficult to scale up. This project will develop a process that enables direct and continuous conversion of flare gas into liquid hydrocarbon fuels. In this process, an initiator was used to facilitate the dissociation of alkanes, which can be converted into liquid hydrocarbons (benzene, toluene, xylenes, mesitylenes, etc) using highly selective zeolite catalysts. This direct conversion will considerably simplify the process, decrease the capital cost and avoid the use of expensive precious metal catalysts and corrosive activators. Therefore, the novel one-step gas-to-liquids process is obviously superior to the current state-of-the-art processes. Moreover, it is expected that the techno-economics of the one-step process will be attractive. This technology has the potential to eliminate flaring for the shale gas industry.

References

[1] http://www.theenergyexchange.co.uk/partner/global-gas-flaring-reduction-ggfr.

480113 Emission Conscious Scheduling of Crude Unloading, Transferring, and Processing for Petroleum Refineries

Abstract:

Emission reduction is becoming increasingly important for petroleum refineries as the increasing concerns of environmental concerns and regulations. Petroleum refinery uses crude oils to produce various fuels such as gasoline, aviation kerosene, diesel, heavy fuel oil, and chemical raw materials such as naphtha and benzene. Crude planning and scheduling is an important part of petroleum refinery due to the potential realization of large cost savings. Cost-effective solution strategies require emission reductions to be addressed from the entire plant point of view, where emission source generations and possible utilizations should be well balanced. But it is a very challenging task due to the large-scale complex manufacturing system, which consists of over 17 major facilities with more than 85 main operation units, and thousands of process streams. The cost-effective emission-reduction strategies should focus on the comprehensively study of material, energy, and information exchanges within the entire manufacturing system, so as to identify the best solutions to the entire plant.

In this paper, a new methodology framework together with a new general scheduling model have been developed for emission conscious crude unloading, transferring, and processing (ECUTP) operations to achieve profitable emission-reduction scheduling (PERS) from the entire system point of view. To identify PERS strategies for refinery plants, a production scheduling model covers both front-end and refinery has to be involved. The key is to optimize the overall material and energy flows to make the plant net profit maximum, meanwhile to ensure emission source generations and utilizations to be smartly balanced. Other manufacturing constraints, such as operation specifications and inventory limits, should also be satisfied as well. Major air emissions from refineries: such as CO_2, volatile organic compounds (VOC), nitrogen oxides (NO_X), and particulate matters (PM) are characterized and quantified.

The scheduling model is a large-scale mixed integer nonlinear programming problem (MINLP). The efficacy of the proposed emission-considered scheduling model has been demonstrated by different case studies. PERS thrusts target on emission source reductions in a profitable and systematic way, which are economically attractive, environmentally benign, and technologically viable for petroleum refineries. It may also help emission generators (not just limited to petroleum refineries) to proactively and systematically reduce their emissions to meet increasingly strict economic and environmental challenges.

480023 Conceptual Design and Cost Estimation of an Offshore Autonomous Once-through Gas-to-Liquid Process Combined with Ammonia Synthesis

Abstract:

Conceptual design and cost estimation of an offshore autonomous once-through gas-to-liquid process combined with ammonia synthesis

A novel process concept is proposed for converting natural gas to liquid Fischer-Tropsch products and ammonia (Figure1). An autothermal reformer with enriched air as oxidant is applied for synthesis gas (syngas) production, and because of the inert nitrogen a once-through Fischer-Tropsch synthesis is the preferred option. In order to maximize the syngas conversion and the production of heavy hydrocarbons, a staged reactor path with distributed hydrogen feed and product withdraw is proposed. The hydrogen is produced by steam methane reforming in a heat exchange reformer (gas heated reformer), heat integrated with the hot effluent stream from the autothermal reformer. The nitrogen needed for ammonia synthesis is produced from air membrane. Tail gas from the last Fischer-Tropsch stage is sent to a gas turbine for power production. The hot exhaust gas from the gas turbine is used for natural gas preheating. The process is autonomous in the sense that it is self-sufficient with power and water, and therefore well suited for production in remote locations such as a floating production unit. The process concept is simple and inexpensive since cryogenic air separation and fired heaters are not required. For the Fisher-Tropsch synthesis microchannel reactors are considered. Cost evaluation is carried out to find out the economic feasibility of the process.

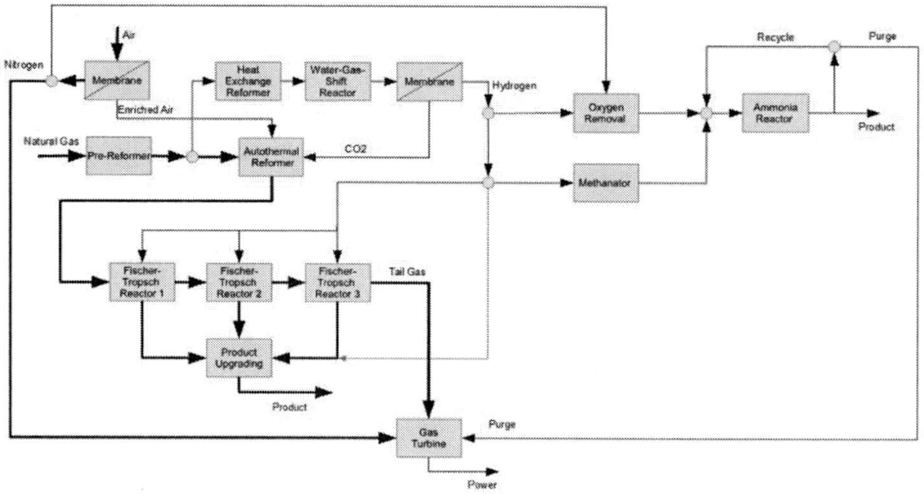

Figure 1: Block flow diagram of the proposed process concept; water and steam are not shown

480897 Study of Liquid Dispersion in Winpak Modular Catalytic Structured Packing: Experiments Vs. CFD Simulations

Abstract:

The response technique was employed to investigate the axial and radial distribution of the liquid phase in Winpak modular catalytic structured packing which was named Winpak-C. Residence time distribution (RTD) experiments were performed at ambient temperature and atmospheric pressure. The axial and radial dispersion coefficients were obtained by solving a two-dimensional dispersion model analytically from the RTD curves. Based on proper assumption, a three-dimensional model of reaction elements in MCSPs was established to determine the dispersion characteristics with the use of computational fluid dynamics (CFD). Porous flow was introduced by using a specification of an extra body force to account for liquid flow through the reaction element. These measured or simulated dispersion coefficients enable the reliable and predictive design and construction of scale-up models.

480908 Process Intensification Toolkit for Mass and Heat Transfer in Rotating Packed Bed

Abstract:

Process intensification (PI) has been widely applied to improve process efficiency on areas such as process capacity and yield, energy consumption, process safety and cost, equipment size, carbon footprint and sustainability, etc. Because of strong shear force that splits liquid into numerous droplets and thin films under a high gravity field, rotating packed beds (RPB) can dramatically increase the mass and heat transfer between gas and liquid. Many research studies [1-3] and industry installations confirm that RPB can reduce equipment volume by 1 to 2 orders of magnitude as compared to conventional packed beds.

With the aim of developing a PI toolkit to support process simulation, research, development, and design of intensified absorption and stripping processes involving RPB, we investigate the effects of gravitational force on gas-liquid interficial areas, mass transfer coefficients, and heat transfer coefficients. The improved understanding on the relationships between gravity and mass and heat transfers allows us to develop a generalied process simulation model for RPB that supports industry's efforts to develop intensified absorption and stripping processes. Examples in the use of the simulation model will be presented.

[1] Yu, C.H., Chen, M.T., Chen, H., Tan, C.S., "Effects of Process Configurations for Combination of Rotating Packed Bed and Packed Bed on CO_2 Capture", Applied Energy, 175, 269-276 (2016).

[2] Yu, C.H., Lin, Y.X., Tan, C.S., "Effects of Inorganic Salts on Absorption of CO_2 and O_2 for Absorbents Containing Diethylenetriamine and Piperazine", Int. J. Greenh. Gas Cont., 30, 118-124 (2014).

[3] Yu, C.H., Cheng, H.H., Tan, C.S., "CO_2 Capture by Alkanolamine Solutions Containing Diethylenetriamine and Piperazine in a Rotating Packed Bed", Int. J. Greenh. Gas Cont., 9, 136-147 (2012).

The Podbielniak® Centrifugal - Applications in Centrifugation

Timothy J. Pomaville
B&P Littleford, LLC
Saginaw, MI, USA

Abstract

Through continuing research and laboratory testing, the Podbielniak® Centrifugal, widely known as the Pod®, has proven itself time and again over broadly differing process applications. Some of the most renowned applications for the Pod, including the streamlining of the penicillin production process, the extraction and concentration of metal ions, as well as many applications in the chemical, petroleum and other industries are described herein along with a case in point to accompany each application. Though there is no new research presented here, recent testing has brought about a shift toward new research of solid phase separation applications to improve upon existing applications. After nearly a century, the Pod continues to have unrealized potential.

A brief discussion of operational theory is also provided in simple form to acquaint the reader with the operational value of the Pod Centrifuge in industry.

Keywords: Pharmaceutical, Chemical, Coke By-products, Hydrometallurgy, Petroleum, Petrochemical, Washing, Separation, Minerals, Mining, Fermented Broths, Solids Handling

Introduction

During the early 1950s, the Pod contactor saw its first, and arguably, one of its most prolific applications take hold within the pharmaceutical industry for the concentration and refining of penicillin. Since this time, the majority of the world's penicillin has been extracted from fermentation broths using the Pod contactor. Over the years, the Pod's numerous applications emerged in petroleum, soap, chemical and metallurgy industries as well. Within the petroleum industry, Pods have been used to replace gravity towers with comparable capacities, greater efficiencies and greater operational flexibility. The description that follows is a simplified explanation intended to lift the shroud of mystery and ignite the reader's imagination with past, present and future applications.

Figure 1: Pod Contactor

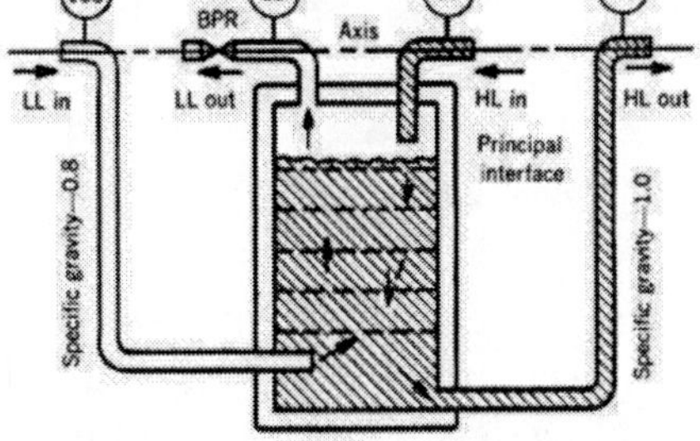

Figure 2: Tower Analogy of a Centrifugal Extractor

Principal of Operation

From an operational point of view, centrifugal force induces the counter-current flow of two liquid phases. The key to controlling an extraction or separation is through control of the principal interface. This is most readily visualized using tower analogy, a comparison to perforated plate, bubble tray or down-comer gravity column to better understand the most important features of the Pod's operation. If the back pressure is increased on the light liquid out stream, the interface will move away from the exit conduit, allowing for more hold-up of the lighter phase within the Pod. In this way, the Pod works the same way a gravity column does, but the centrifugal force generated can be 100s of times that of a gravity column. Like any other liquid/liquid extraction unit operation, the Pod is then characterized by the number of theoretical stages calculated for a given application.

What's Physically Happening Inside?

A – Heavy liquid out (HLO) - leaves the rotor after being centrifuged to the outer rim.

B – Light liquid out (LLO) – leaves via the shaft after being displaced inward.

C – Light liquid in (LLI) enters the rotor slightly inboard of the rim and the HLO.

D – Heavy liquid in (HLI) enters an outboard from the shaft. It is then centrifuged outboard to the rim, displacing the light liquid, inducing the counter-current flows.

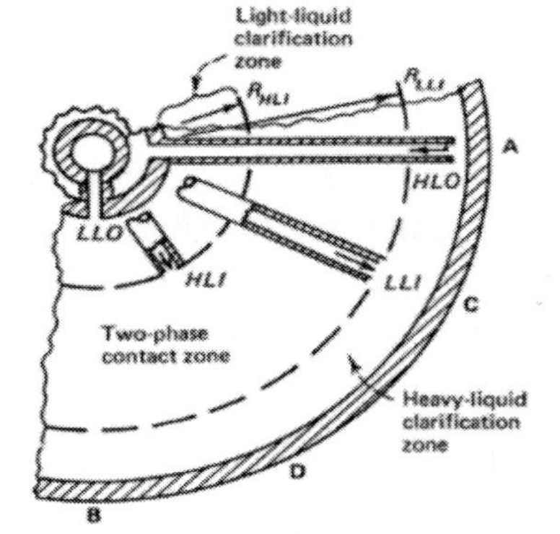

Figure 3: Cross-section of a Pod rotor

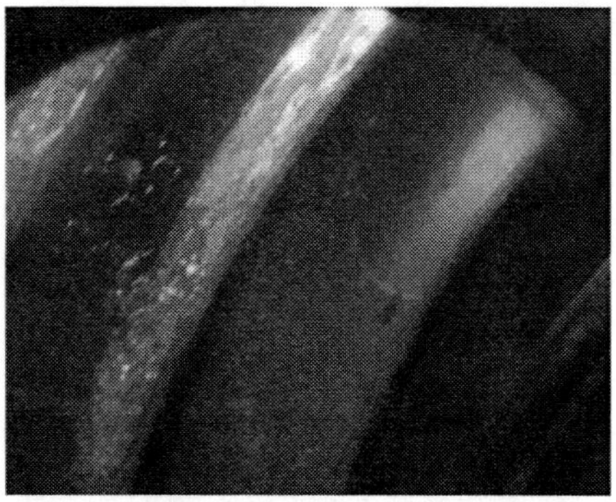
Figure 4: Window rotor Pod in operation

The internal fluid dispersion shown in Fig. 3 illustrates the above concept. In general, to observe the intended effects of a Pod, there must be a differential in the specific gravity of at least 0.01. This difference in specific gravity allows for the manipulation of the interface through adjustment of pressure differential. A continuous phase can be chosen for the whole rotor if the process dictates the need. With the interface near the shaft, the dispersed phase will be the light liquid or light key. Conversely, if the interface is near the rim, the dispersed phase will be the heavy liquid or heavy key. Selection of perforated element bands, along with careful positioning of the fluid entry points, affords flexibility from one application to another.

For simplicity and brevity, imagine the washing operation of kerosene oil with water. This is a classic example which is used repeatedly for demonstration purposes. Water, the heavier liquid is introduced to the Pod to lubricate and keep cool the hermetic mechanical seals. The rotor is started, and kerosene oil is introduced to the system. A back-pressure is then applied to the light liquid out (LLO) stream. Heavy liquid out (HLO) increases momentarily as a portion of the volume of water within the rotor is displaced. At this time, the principal interface begins to move outboard, away from the shaft. After a short time, the system stabilizes and reaches a steady state. Light liquid enters near the rim, leading rotation and passing through the heavy liquid of the heavy liquid clarification zone by way of displacement due to the specific gravity difference in the form of droplets. Small droplets (higher surface area) expedite the extraction of impurities from the kerosene oil into the water phase. Simultaneously, heavy liquid enters

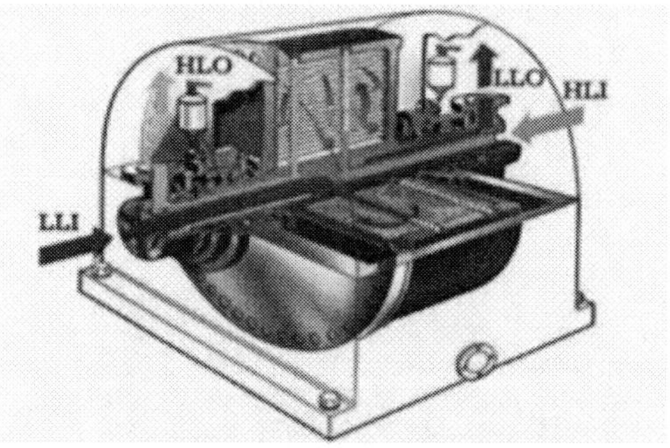

Figure 5: Flow through a Pod

the system just outboard of the shaft, lagging rotation and light liquid clarifying zone. Droplet formation is facilitated by the bands, exposing surface area of the heavy phase, allowing impurities to pass into the water or solvent phase. The water, containing impurities, is sent outboard to exit at the periphery of the rotor (rim).

Pharmaceutical Pod Applications

The Pod's earliest successes were seen in the pharmaceutical industry, as evidenced by high volume production of antibiotics such as penicillin and other fermentation products. In fact, problems that were particular to large-scale production of penicillin heavily influenced the development of the Pod. As the pharmaceutical industry grew, the Pod evolved to accommodate growing fermentation capacities until a full line of standardized Pods were developed with features that emphasize easy maintenance and cleaning. Certainly, these developments have been utilized in other industrial markets, but the Pod has matured through the pharmaceutical industry and maintains a significant role in its extraction processes.

Fundamental to the maximum recovery of unstable products from low pH, non-refrigerated broths, as in the case of penicillin production is a swift contacting time from the fermenter to the finished product. The Pod lent itself well to solving these problems with commercial scale pharmaceutical production that were previously not possible. Since the potency of the acidulated broths is time dependent, the fraction-of-a-minute holdup time permitted maximum recovery of product potency despite its relatively short half-life. Additionally, many broths that show extreme tendencies to emulsify due to the many active agents added to their respective fermentation processes can be recovered in a Pod.

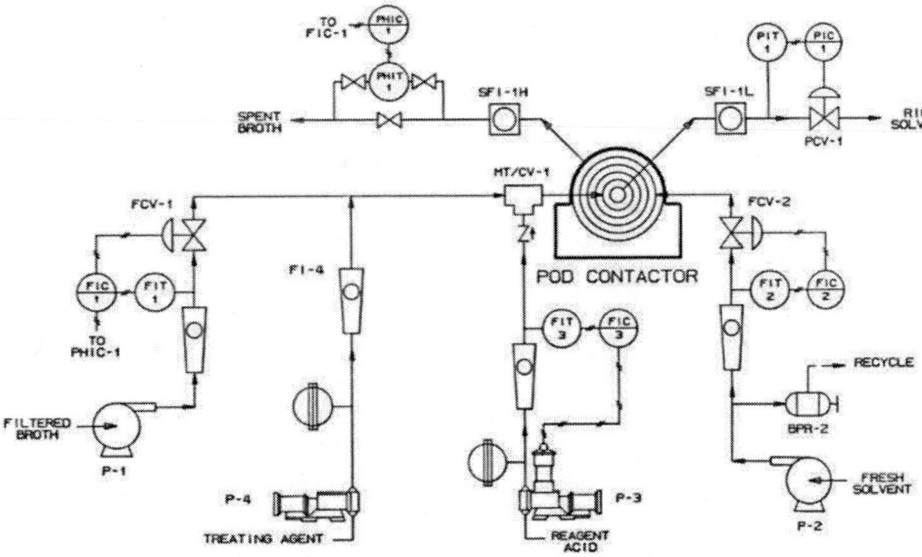

Figure 6: Primary Antibiotic Extraction

Although it is of course process dependent, Pods have seen upwards of five theoretical stages within one unit, contributing to 98% or better extraction efficiency in a single step. Selective solvent to broth ratios were minimized, permitting maximum concentration of products and effectively reducing solvent inventories. At the same time, high G-forces imparted by the Pod permitted maximum clarity to the product streams.

These advantages specific to the penicillin process became widely used in industry for the production of antibiotics such as Bacitracin, Cloromycetin, Kanamycin, Tetracyclin, Estrogens, Steroids, Cortisone, Ephedrine, Opiates, a variety of vitamins and many others. Large scale Pods were subsequently developed and standardized for the production of pharmaceuticals to process greater volumes of fermented broth. The stream arrangement shown Fig. 4 is the most common in pharmaceutical applications. Many pharmaceutical producers using Pods practice counter-current extraction on one or two Pods working in series with only slight variations that depend on broth quality at a given time. In some cases, producers have used special techniques tailored in-house to meet their own, custom applications.

Petrochemical Pod Applications

Waste streams from refineries containing fractions of naphthenic acids can be removed by treating with caustic. The sodium naphthenates formed in the process can be easily disposed or reused to regenerate naphthenic acids for further use. A simplified flow diagram for the recovery of naphthenic acids from refinery waste streams is illustrated below in Fig. 6. Spent caustic is mixed in-line with a strong inorganic acid and resulting in a weaker organic acid. The emulsified mixture enters the Pod to be washed with water for the removal of impurities. Recoveries of naphthenic acid in excess of 99% have been observed. The salt water, with acceptable impurities, can be sent to the sewer. The pH of the salt water stream can be continuously analyzed to easily automate changes in the feed ratios.

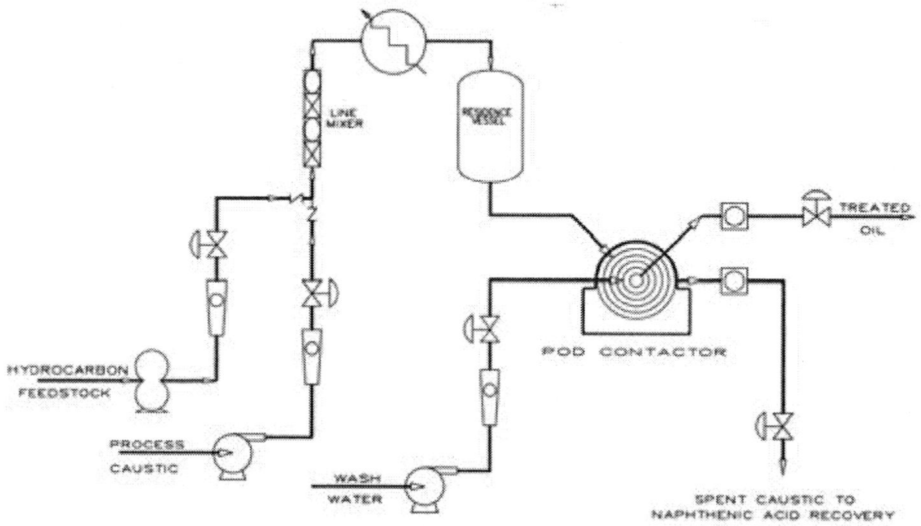

Figure 7: Caustic Treat/ Water Wash to Remove Nap Acids

Solvent extraction from lubricating oils was proven decades ago in the petroleum industry. A single pass through the Pod using a selective solvent produces a raffinate of outstanding quality. Undesirable components are removed and rejected to the extract stream where they can be stripped and reused for subsequent extraction. As compared to a commercial perforated tray column, the Pod requires only 1/45 of the holdup with reduced off spec production when changing from one grade of lube oil to another.

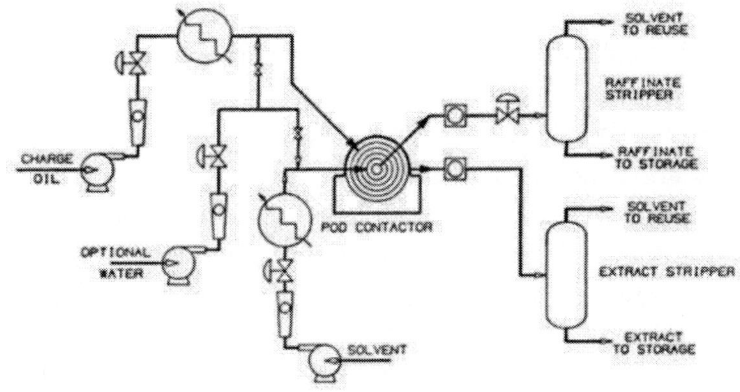

Figure 8: Simplified Solvent Lube Oil Extraction

Within the petrochemical industry, Pods can also be used to extract aromatics and coke-forming compounds in connection with catalytic cracking operations. Cycle stock would be used to raise crude charge stock capacity, catalytic cracker charge capacity and percent yield of gasoline. Use of a Pod as an extractor reduces coke burning requirements, increasing gasoline yield and reduces catalytic cracker investments.

The advantages are numerous and similar to those previously mentioned for other applications. Controlled reactions, solvent extractions and separations are all accomplished with a Pod in the continuous sulfonation application, for example. Shorter processing times for the separation of sludge from an oil/sulfonate mixture are accomplished in minutes as opposed to hours. Light colored sulfonates are produced due to the elimination or overreacting or "burning." The Pod makes the best use of oleum with no side reactions, such as unstable or off color products, to name a few benefits.

Hydrometallurgical Pod Applications

Pods have been used extensively for the extraction, concentration and purification of molybdenum, zirconium, vanadium and titanium metal complexes, as well as rare earths such as thorium and cesium through the use of selective solvents. Pods are especially useful in recovery of regenerative complex liquid ion exchange mediums that are dissolved in solvents immiscible with aqueous solutions. In fact, the bulk of practical solvent extractions involve complex formations as a way of influencing the distribution in the system.

In hydrometallurgical applications, an aqueous solution containing the desired metal is contacted with a suitable solvent. The metal, in some form, will distribute between the aqueous and organic phases. To make the metal favor the organic phase, an extractant or reagent is often used which can react with the metal to form complexes. These metal complexes are the predominant species that are extracted into the organic phase. For hydrated metal ions existing in aqueous solution which cannot be extracted to a significant extent, organic solvents are reacted to replace the water of hydration. This is accomplished by combining the metal ions with ions of an opposite charge to form a neutral metal species. The larger and more hydrophobic the resulting species, the better the extraction will be.

An especially useful application for Pods is in the extraction of rare earths from sulfate, nitrate or chloride solutions. Though some of these processes have certainly evolved, a 15 second pass through a Pod has demonstrated that at 500 GPM of pregnant sulfuric acid leach liquor using 80 GPM organic solvent can yield up to 0.8 gpl of rare earth metals (loading step). A single pass through a smaller Pod (stripping step) transfers rare earth salts into an ammonium nitrate solution for the recovery of these various rare earth metals by precipitation into an appropriate solution. The organic solvent is then recovered via scrubbing and recycled back to the first step to complete the loop.

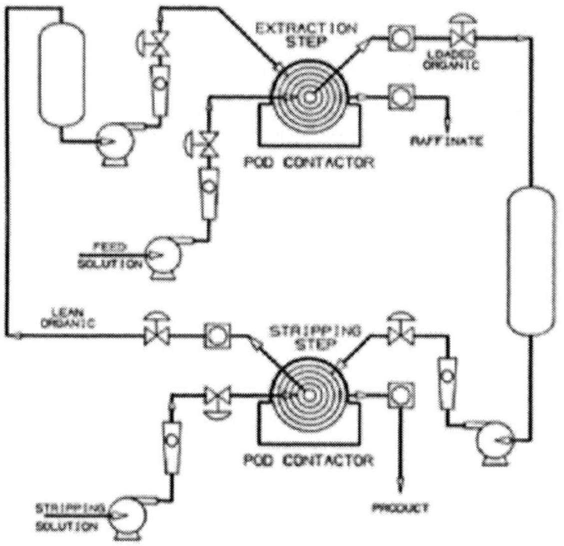

Figure 9: Simplified Hydrometallurgical Liquid Ion Exchange

A conventional mixer-settler plant would need to use 60,000 gallons of solvent (5-10% solution of Amine complexing agents), while a Pod would require only 3,000 gallons to accomplish the same task. This amounts to a 20 fold decrease in solvent inventory. With its multiple theoretical stages, solvent extraction in one Pod has been seen to leave only one ppm of rare earth metal in the waste liquor. Automated instrumentation assures close control of the process with minimal operator interaction. The control system is actuated by changes in the flow of pregnant liquor and ratio controllers adjust other flow rates accordingly.

Pods, here again, have the advantage of up to 5 theoretical stages for the extraction of metals and the use of centrifugal force to disengage the phases. Because of the short hold-up time within a Pod, they can be used to separate two or more metals contained in a given solution when the reaction rate for one metal is considerably shorter than another as opposed to columns or mixer settlers. Varying the RPMs or throughput can permit improved selectivity in the separation of a desired metal

Upcoming Research

A significant interest in many Pod applications arises when microscopic solids are introduced to the system. There are many existing Pod applications that would see improvement through shedding of solids from the internals, especially when shutting down the process is a drawback to the user. A solution will require extensive research that may need to be evaluated for several months to arrive at any definitive conclusions. Some suggestions have been proposed to keep solids from depositing, but much of the work has yet to be attempted. Promising results in testing have recently provoked a new interest in this behavior as it relates to solids in the Pod. Using a system of two immiscible liquids, one of them containing particularly challenging solids, research to smooth the progress of existing and promising new applications are being thoroughly investigated.

Conclusion

The Pod centrifugal contactor has extraordinary processing capabilities and a well-built mechanical design. They have been used broadly and successfully in many applications and continue to have untapped potential in many other untested applications. For these untested applications, it may be necessary to conduct laboratory testing for specific materials or process requirements to properly evaluate operating conditions and performance. While it is possible to test at off-site facilities, feasibility and pilot testing in the Saginaw, Michigan facility is available for a wide variety of different process configurations along with all of the auxiliary equipment required of commercially standardized models.

Summary of Benefits

The performance of the Pod is unique in its applications. It has become generally accepted that a single unit provides up to five theoretical stages of counter-current contact. One Pod has sufficiently high efficiency to achieve phase separation and counter-current washing of effluent product streams. The largest Pod available today can achieve up to 600 GPM of total combined flow. They are compact, easy to operate and readily cleaned and transported. Materials of construction are stainless steel and other corrosion resistant alloys. Liquid holdup within a Pod ranges from 1-260 gallons depending on the model. Liquids can be passed through a Pod in a fraction of a minute, and if longer residence times are required, liquid flows can be easily adjusted or recycled. Steady state conditions are reached quickly upon start-up and small continuous runs can be completed quickly and effectively with negligible transitional fractions. Pods are inherently pressure tight, closed systems, handling up to 250 psig and up to 600 psig by special design. In addition to being pressure tight, they are suitable for operation from below 0°F - 250°F and 450°F by special design. They are low maintenance with exceptionally long life expectancies.

References

[1] Davies, G.R., Kaiser, H.R., Todd, D.B., "Controlled Centrifugal Separation of Liquids": Baker Perkins Inc., Saginaw, Michigan

[2] Todd, David B., Davies, Gordon R., "Performance of Centrifugal Extractors": International Solvent Extraction Conference, Lyon, Vol.3, 2380 – 2398, September 1974

[3] Todd, David B., Davies, Gordon R., "Centrifugal Pharmaceutical Extractions": Filtration & Separation, November/December 1973

[4] Anderson, S.O., Ashbrook, A., etal, "Solvent Extraction in Process Metallurgy": University of Waterloo and Mines Branch, September 1971

[5] Todd, David B., Podbielniak, W.J., "Advances in Centrifugal Extraction": Chemical Engineering Progress, May 1965

2017 AIChE Spring Meeting Presentation
Paper 99c, Abstract #481626

CRITICAL PARAMETERS FOR DESIGN, TROUBLESHOOTING AND OPTIMIZATION OF AN ETHERIFICATION PROCESS

Minghua Ye, Sergio Mohedas, Brian Peng, and Chih Ning Kuan

March 28, 2017

CHEMISTRY THAT MATTERS™

REACTION: REVERSIBLE REACTION

Low temperature synthesis: **Tertiary Olefins + Alcohols → Ethers**

High temperature back cracking: **Ethers → Tertiary Olefins + Alcohol**

Application:

- Convert low value methanol and ethanol to high value fuel
- Improve octane number for gasoline fuels
- Produce high purity isobutene (incorporate the MTBE back-cracking process)

Typical ethers:

- MTBE (Methyl Tertiary Butyl Ether)
- TAME (Tertiary Amyl Methyl Ether)
- ETBE (Ethyl Tertiary-Butyl Ether)

No. 1

CONCEPTUAL ETHERIFICATION PROCESS

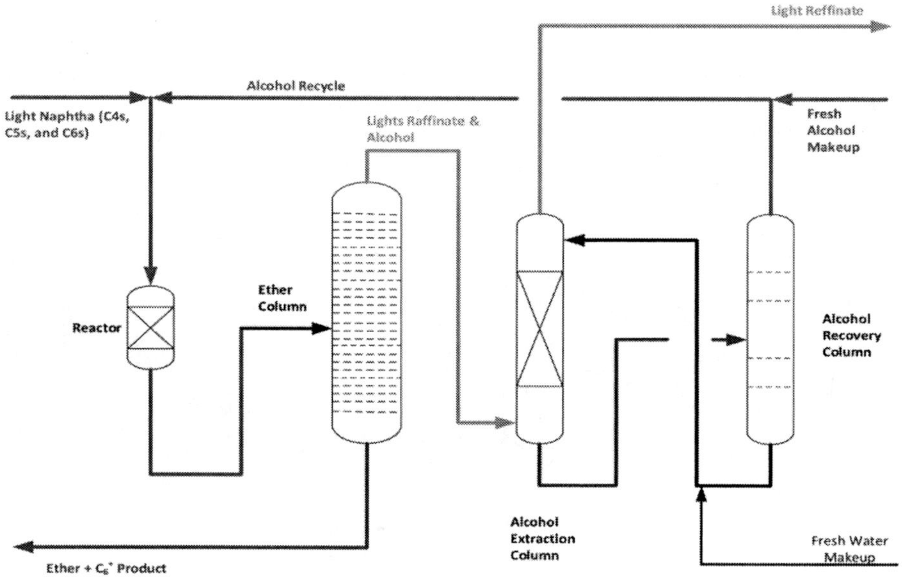

No. 2

ADDING FEED PRETREATMENT UNIT TO REMOVE CONTAMINANT (PROTECT CATALYST)

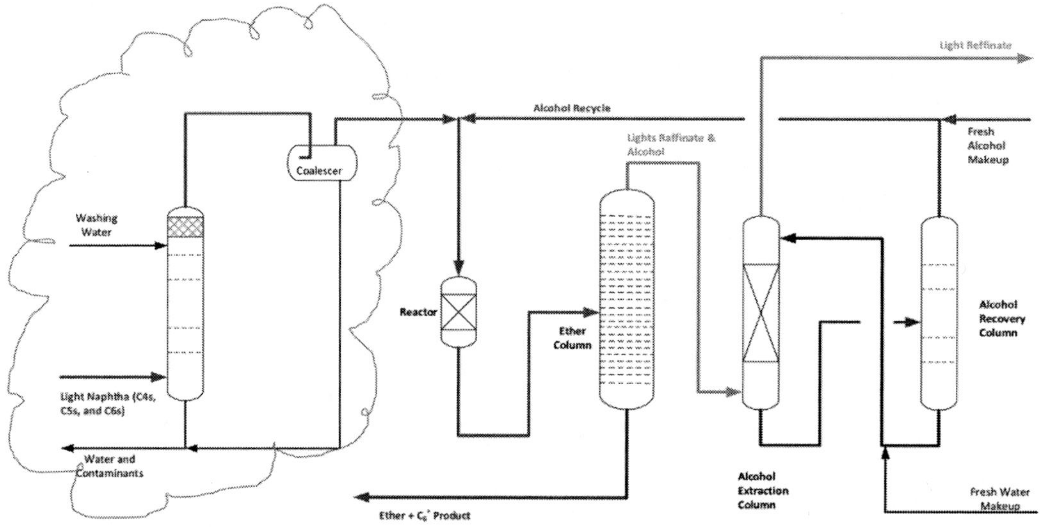

No. 3

INCORPORATION OF REACTIVE DISTILLATION TECHNIQUE TO IMPROVE OVERALL CONVERSION

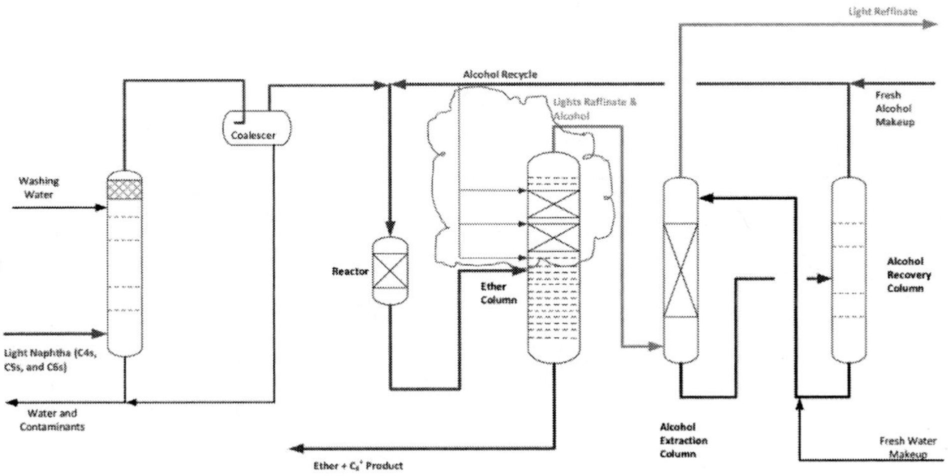

No. 4

CRITICAL PARAMETERS FOR PROCESS DESIGN, TROUBLESHOOTING AND OPTIMIZATION

- Molar ratio of alcohol to reactive olefin
- Etherification reactor operating conditions
- Azeotrope / reactive distillation
- Effective recovery of alcohol

No. 5

KEY POINTS

- Feed quality
- Reactor conversion / selectivity
- Reactive distillation column conversion / selectivity
- Overall conversion / selectivity
- Alcohol consumption
- Utility and energy consumption

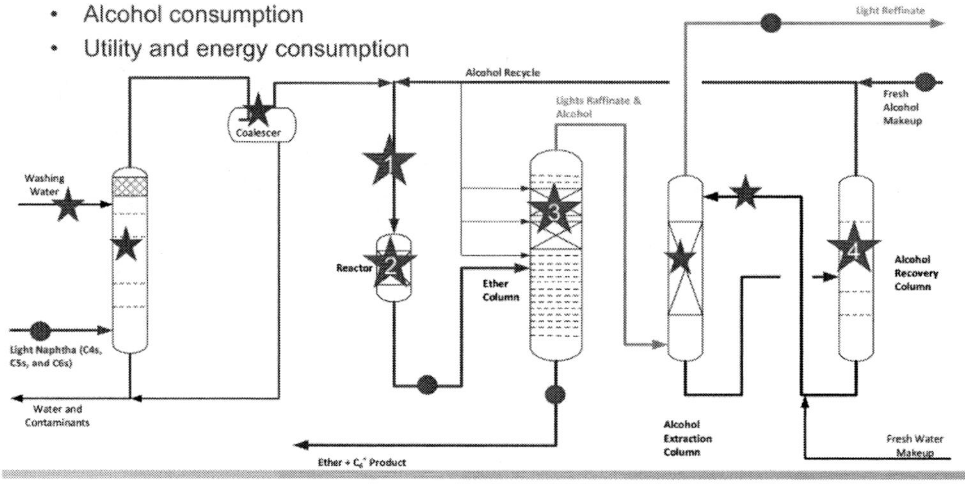

No. 6

PROTECT CATALYST: FEED PRETREATMENT

Catalyst Deactivation

Permanent:

- Nitrogen compounds (such as acetonitrile)
- Metal Ions (Ca^{++}, Mg^{++}, Fe^{+++}, NH_4^+)
- Catalyst fouling by polymer forming agents (such as butadiene, pentadienes)

Temporary: High water content in feed (resulting in the loss of sulfonic acid groups)
Water may react with tertiary olefin: $iC_4H_8 + H_2O \rightarrow$ TBA (Tertiary Butyl Alcohol)

Impact:

- Reduce catalyst activity
- Shorten the catalyst life
- Increase reactive distillation column loading and operation

Feedstocks:

- The reactive olefins vary from different operating mode of upstream units.
- The contents of impurities must meet specification set by catalyst vendor.

No. 7

MOLAR RATIO OF ALCOHOL TO REACTIVE OLEFIN

$$R = \frac{Alcohol\ Molar\ Flow\ Rate}{Total\ Molar\ Flow\ Rate\ of\ Reactive\ Olefins}$$

Reactive Olefins are the tertiary olefins.

High alcohol ratio can result in alcohol dehydration, for example:

$CH_3OH + CH_3OH = CH_3\text{-}O\text{-}CH_3 + H_2O$

DME formation must be limited especially because:

- it is a very volatile compound and will impact the RVP, an important index of gasoline
- It is a poison in alkylation process

Low alcohol ratio can result in:

- Low reactive olefin conversion
- Dimerization of reactive olefins: 2 $i\text{-}C_4H_8 \rightarrow$ dimer
- Increment of reactive distillation column loading

MAIN REACTIONS IN THE ETHERIFICATION PROCESS

Methanol + tertiary olefins to form ethers

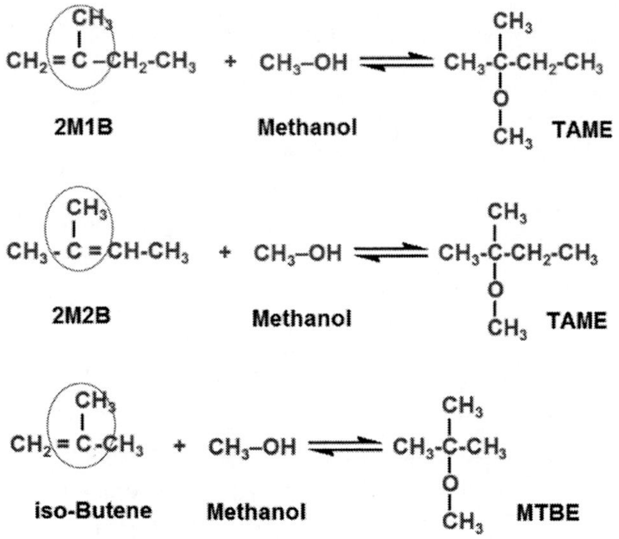

REACTIVE OLEFINS – TERTIARY OLEFINS

C4 Olefin	isobutene		+ CH_3OH ===	Methyl tertiary butyl ether (MTBE)
C5 Olefin	2-methyl-2-butane	2-methyl-1-butane	+ CH_3OH ===	2-methyl-2-methoxy butane (TAME)
C6 Olefin	2-methyl-1-pentene	2-methyl-2-pentene	+ CH_3OH ===	2-methyl-2-methoxy pentane
	cis-3-methyl-2-pentene	trans-2-methyl-2-pentene	+ CH_3OH ===	3-methyl-3-methoxy pentane
	2-ethyl-1-butene			
	2,3-dimethyl-1-butene	2,3-dimethyl-2-butene	+ CH_3OH ===	2,3-dimethyl-2-methoxybutane
	1-methyl-cyclopentene		+ CH_3OH ===	1-methyl-1-methoxy cyclopentane
C7 Olefin	2-methyl-1-hexene	2-methyl-2-hexene	+ CH_3OH ===	2-methyl-2-methoxy hexane
	cis-3-methyl-2-hexene	trans-3-methyl-2-hexene	+ CH_3OH ===	3-methyl-3-methoxy hexane
	cis-3-methyl-3-hexene	trans-3-methyl-3-hexene		
	2-ethyl-2-pentene			
	2-methyl-1-pentene	2-methyl-2-pentane	+ CH_3OH ===	2,3-dimethyl-2-methoxy pentane
	cis-3.4-dimethyl-2-pentene	trans-23,4-dimethyl-2-pentene	+ CH_3OH ===	2,3-dimethyl-3-methoxy pentane
	2-ethyl-3-methyl-1-butene			
	2,4-dimethyl-1-pentene	2,4-dimethyl-2-pentene	+ CH_3OH ===	2,4-dimethyl-2-methoxy pentane
	3-ethyl-2-pentene		+ CH_3OH ===	3-ethyl-3-methoxy pentane
	2,3,3-trimethyl-1-butene		+ CH_3OH ===	2,3,3-trimethyl-2-methoxy butane
	1-ethyl cyclopentene		+ CH_3OH ===	1-ethyl-1-methyoxy cyclopentane
	1,2-dimethyl cyclopentene	1,5-dimethyl cyclopentene	+ CH_3OH ===	1,2-dimethyl-1-methoxy cyclopentane

Ref: Halim Hamid Mohammad Ashraf Ali, Handbook of MTBE and Other Gasoline Oxygenates, Marcel Dekker, Inc, pp 269

No. 10

SIDE REACTIONS

(1) Methanol Dehydration

$2\ CH_3OH \rightarrow DME + H_2O$

Kinetically favored by high temperature, high concentration of methanol

DME is a very volatile compound that leaves the process with C_4 stream

DME is a poison in alkylation

(2) Dimerization of isobutene to give C_8 hydrocarbons

$2\ iC_4H_8 \rightarrow$ Dimer (C_8H_{16})

Favored by low methanol concentration

Leave the process unit with MTBE, decreasing MTBE purity.

(3) Isobutylene Hydration for tert-Butyl Alcohol

$i\text{-}C_4H_8 + H_2O \rightarrow TBA$

Reaction is promoted by the quantity of water present

Affecting MTBE purity, catalyst activity, and overall conversion

(4) Normal butylene etherification:

$N\text{-}C_4H_8 + CH_3OH \leftrightarrow$ 2-methoxy butane

Kinetically limited, very low conversion, favored at high temperature

2-methoxy butane have much lower octane number than MTBE

Cannot be separated from MTBE

No. 11

ETHERIFICATION REACTOR OPERATING TEMPERATURE

Exothermic and equilibrium limited, liquid phase reaction
Fixed bed reactor is commonly used in industry
Lower T: Require more catalyst, longer catalyst life, lower DME and Dimer formation
High T: Require less catalyst with shorter life, higher DME and Dimer formation

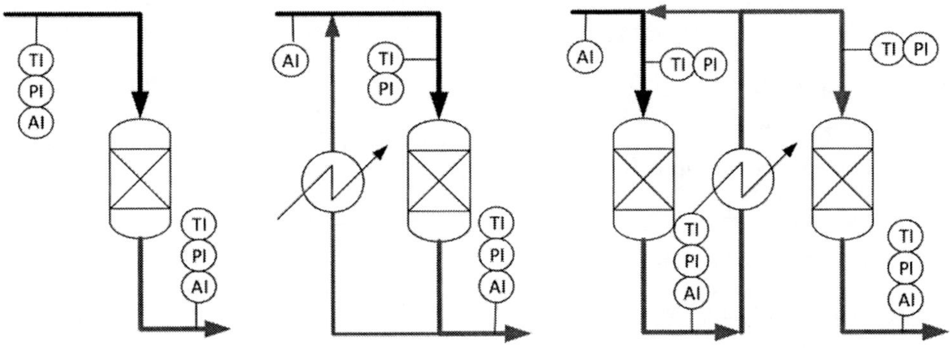

Single Reactor | Single Reactor with Cooling Recycle | Two Reactors with Cooler and/or Recycle

No. 12

ETHERIFICATION REACTOR OPERATING PRESSURE

The operating pressure depends on the reactor type, operating temperature, and reactor outlet pressure required to deliver the product to downstream units.

- Liquid Phase Reaction (no vaporization in the reactor):
  P_{out} > bubble point pressure at reactor exit temperature.
  P_{out} > down stream separation unit set pressure + pipeline pressure loss
  $P_{in} = \Delta P_{bed} + P_{out}$

- Vaporization Fixed Bed Reactor (reaction heat is removed by partially vaporizing the liquid mixture)
  P_{out} < bubble point pressure at reactor exit temperature
  P_{out} > down stream separation unit set pressure + pipeline pressure loss
  $P_{in} = \Delta P_{bed} + P_{out}$

No. 13

REACTIVE DISTILLATION (ACHIEVING HIGH ETHERIFICATION CONVERSION)

Thermodynamics:

- highly non-ideal VLE and LLE for HC, alcohol, and ether mixture system
- Low temperature azeotropes of alcohol-HC form in the reactive distillation columns.

Operating parameters

- Alcohol injection – Promote further etherification in the catalyst bed zone
- Overhead HC + alcohol, azeotrope with lower temperature boiling point
- Column operating T, P, and reflux rate

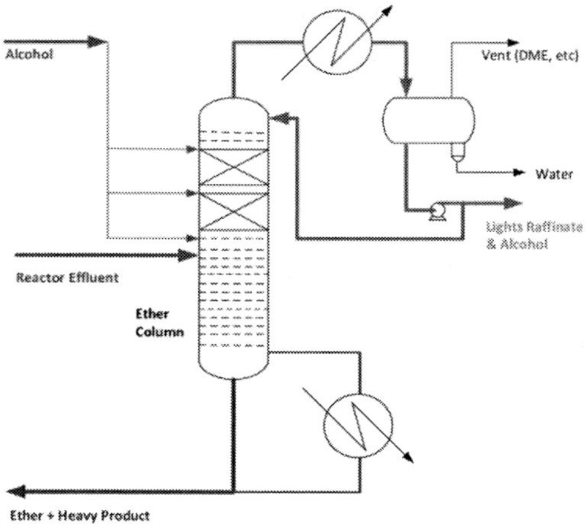

No. 14

ISOBUTENE + METHANOL + MTBE BINARY VLE DIAGRAMS

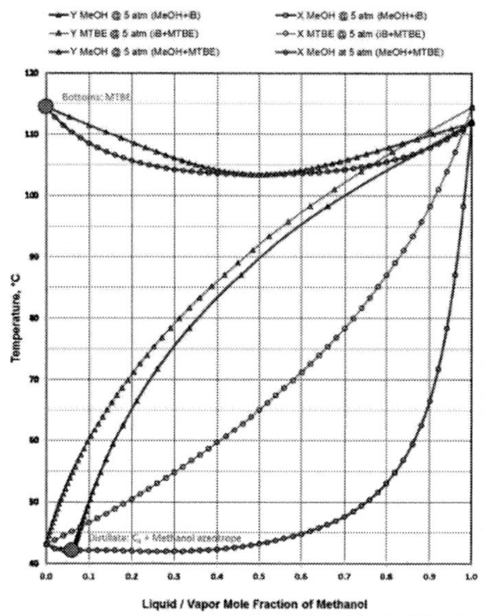

iso-Butene, Methanol, MTBE Binary

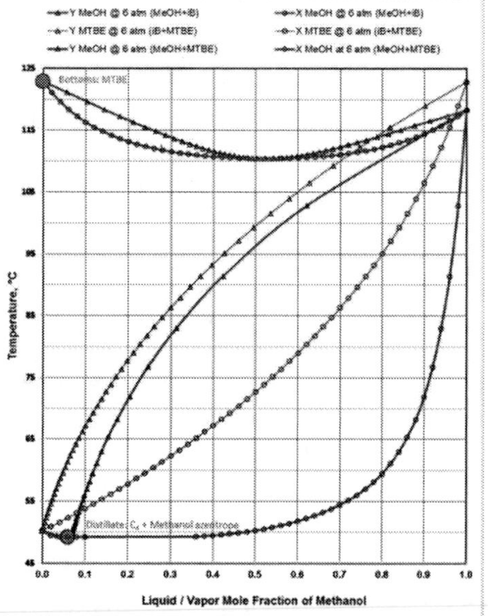

iso-Butene, Methanol, MTBE Binary

No. 15

2-METHYL PENTENE-1 + METHANOL + MTPE BINARY VLE DIAGRAMS

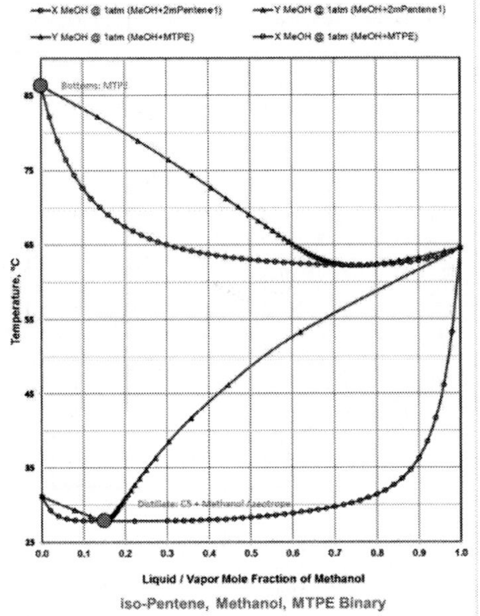

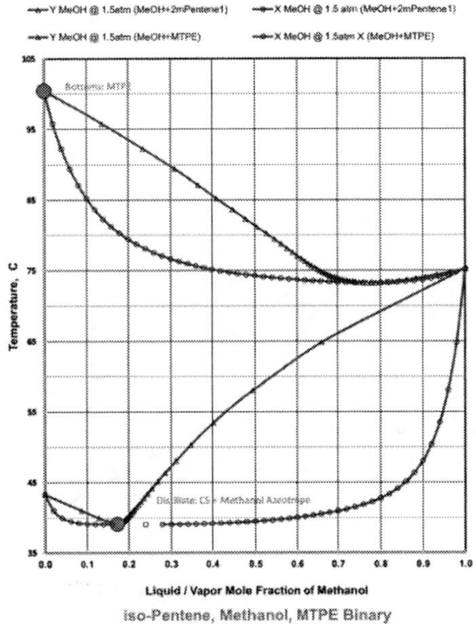

ALCOHOL EXTRACTION

Excess alcohol recovery:

- Liquid extraction of alcohol with water
- Provide sufficient theoretical stages (number of trays or the height of structured packing) to minimize washing water amount
- Use optimal amount of washing water

Excess water could results in:

- Larger extraction column size
- More energy to separate alcohol from the aqueous solution
- Larger size for the equipment in the alcohol separation unit (distillation column, condenser, reboiler, drums, etc.)

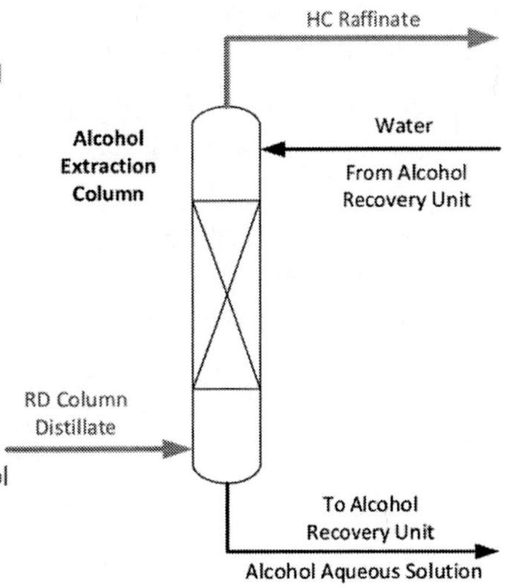

HIGH PURITY ALCOHOL RECOVERY

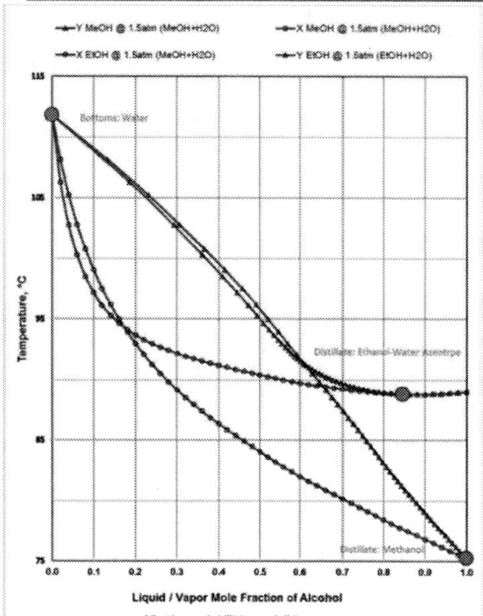

Methanol / Ethanol Binary

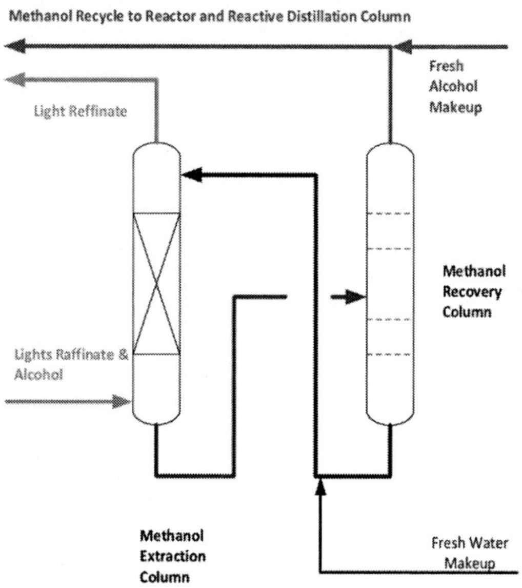

No. 18

REQUIREMENT FOR ALCOHOL PURITY

- The etherification reactor and reactive distillation bed require water content in feed to be less than 1000 ppm. Both the recovered and fresh alcohol need to be monitored to ensure the recycle alcohol quality.
- For methanol recovery unit, it is achievable by distillation to obtain 99.9 wt% methanol.
- For ethanol, additional extractive distillation technique may be applied to obtain high purity ethanol.
- Alcohol content in the bottoms (water) to be well controlled. High alcohol content in water result in
  - ➔ Alcohol loss to raffinate (extraction column overhead)
  - ➔ Alcohol loss to purge stream

No. 19

SUMMARY

The critical parameters for etherification process:

- Feed pretreatment to remove catalyst poisons from feed
- Set optimum molar ratio of alcohol to reactive tertiary olefins prior to reactor
- Optimal etherification reactor operating conditions to compromise reasonable reaction kinetic rate and equilibrium conversion
- Reactive distillation for etherification involves formation of azeotropes of HC-alcohol, optimum alcohol injection amount ensures the etherification reaction takes place at low temperature (bubble point of HC-alcohol azeotrope). Excess methanol will result in high alcohol content in the ether product, and higher energy consumption for alcohol recovery
- Effective recovery of alcohol (extraction and distillation) is important to achieve high energy efficiency, recycle alcohol water content, overall reactive olefin conversion, and the overall unit performance

No. 20

DISCLAIMER

No. 21

480961 Choosing and Using Tollers to Manufacture

Abstract:

Pilot plants are an endangered species. There are a variety of reasons for this, CAPEX limitations, safety concerns, environmental risks, resource limitations, and time constraints. CAPEX is in high demand and short supply. All projects are in competition for limited available capital funds. Regarding safety, it is key to ensure that personnel are properly trained, appropriate equipment is used, and chemical inventories are kept low. Environmentally it is important to contain and properly dispose of chemicals, which requires staff training and adherence to regulations, both federal and state. Developing qualified personnel to operate the unit requires time and training. Project cycle time is paramount and essential to maintain a competitive advantage.

The use of contract manufacturers at the pilot plant level provides a seedy solution to these problems. The majority of the investment has already been made and basic safety issues have been addressed. Only safety issues specific to the project require further assessment. Environmental risks should be discussed in detail with the contract manufacturer since you (the contracting company) own the environmental liability for the project. Lastly, the unit is already staffed with trained operators. Time constraints are now limited to toller schedule and feed logistics.

This presentation discusses

- How to identify tollers
- How to transfer the technology while maintaining IP
- How to implement the technology at the toller
- How to successfully manage the effort.

Case studies of our experiences, some of which went well, others not so well will be discussed

479975 To Toll or Not to Toll: That Is the Question

Abstract:

Globalization is the result of

- the recovery of international production capacity following its devastation during the Second World War;
- the revolution in communications;
- the revolution in transportation.

The first result means that global production is at an all-time high: everybody is producing everything. The communications revolution means that everybody can market globally, cheaply via a Website on the World Wide Web. The transportation revolution means that product moves almost anywhere for almost no cost. The above bullets sum to a tremendous increase in commercial competition since the early 1990's.

One way to confront this increased competition is to reduce process and product "cycle time", where cycle time means the period between invention and commercialization. The shorter cycle time is, the faster product enters a given market. Using contract manufacturing is one method for reducing cycle time in the chemical processing industry.

This presentation discusses the "in and outs" of using contract manufacturing to reduce cycle time in the chemical processing industry. This presentation is the result of twenty-five years' experience at using contract manufacturing to reduce cycle time.

2017 AIChE®

SPRING MEETING & 13TH GLOBAL CONGRESS ON PROCESS SAFETY

March 26–30, 2017
Henry B. Gonzalez Convention Center
San Antonio, TX

17 AIChE
Spring Meeting, San Antonio, TX

13TH GLOBAL CONGRESS ON PROCESS SAFETY

Integrated LPG Upgrading Process Using Divided Wall Columns

Scott Schroeder, Advisian
scott.schroeder@advisian.com

Armen Abazajian, Advisian
armen.abazajian@advisian.com

www.advisian.com

Advisian
WorleyParsons Group

Process Intensification

A philosophy of design that seeks to bring dramatic improvements in the process efficiency, equipment size, or waste reduction through the use of process-intensifying equipment or process-intensifying methods.

Process-intensifying equipment – novel reactors such as micro-channel reactors or rotating packed-bed reactors; novel heat exchangers such as micro-channel heat exchangers

Process-intensifying methods – such as hybrid separations, integration of reaction and separation, integration of heat exchange and separation

Process Intensification

The field of Process Intensification has developed a large tool set of equipment and methods

This tool set includes mature technologies that pre-date the term "Process Intensification" as well as new cutting edge technologies.

Process Intensification - DWC

One traditional method of process intensification is the integration of distillation sequences.

Distillation is the one of the most common separation operations used in the process industry. It accounts of a large portion of the energy usage. Reduction of energy usage has been a key driver in developing integrated distillation sequences.

Divided Wall Columns (DWC) take the concept integrating a distillation sequence to the equipment level.

Conventional Column Sequences

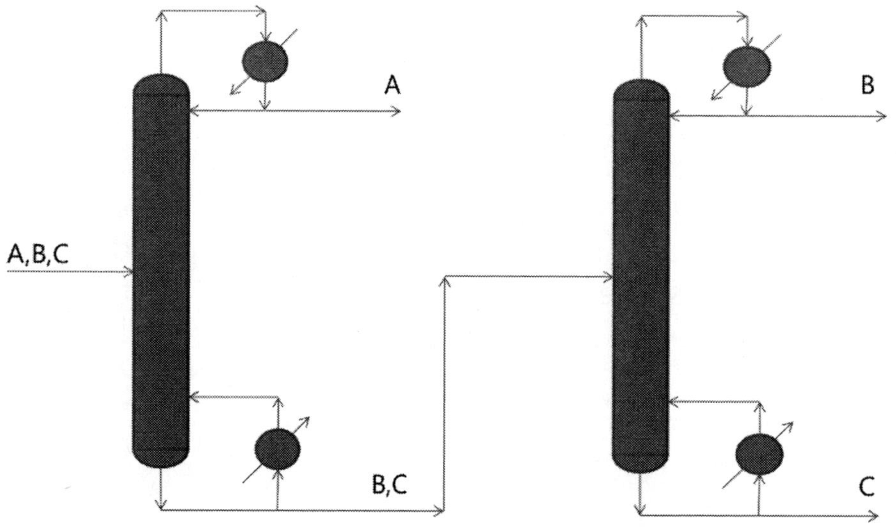

Conventional Column Sequences

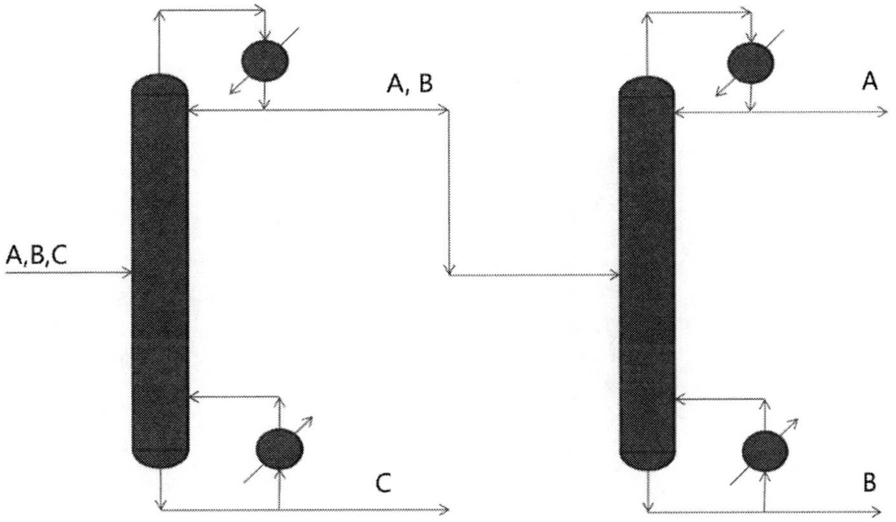

2017 AIChE® SPRING MEETING & 13TH GLOBAL CONGRESS ON PROCESS SAFETY

Advisian / 7

Petlyuk Column

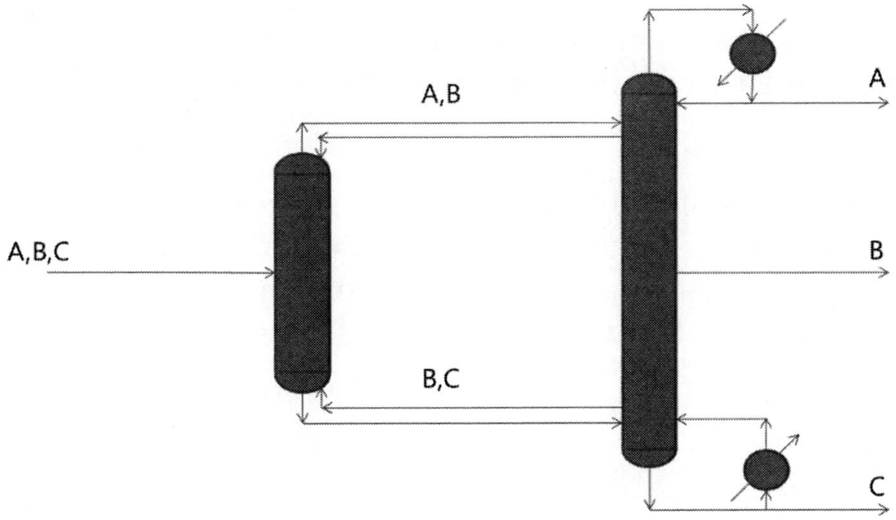

2017 AIChE® SPRING MEETING & 13TH GLOBAL CONGRESS ON PROCESS SAFETY

Advisian / 8

Divided Wall Column (DWC)

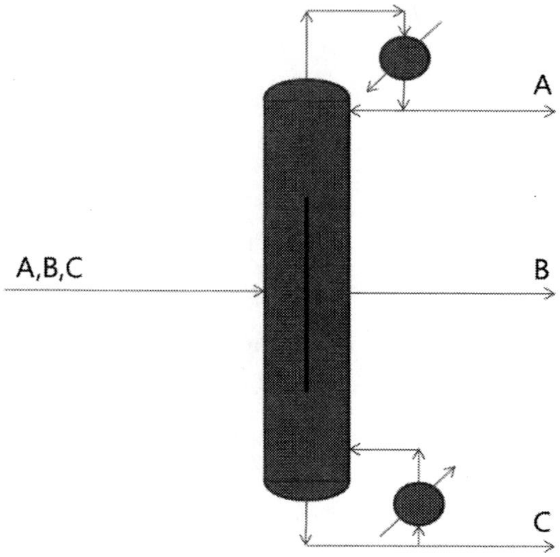

Process Intensification – Reaction Coupling

Another Process Intensification method is to couple separation and reaction to optimize the overall process.

An example would be using a reaction to convert a mixture with low relative volatility that would be difficult to separate with distillation into a mixture with that has high relative volatility that will separate much easier by distillation.

LPG Upgrading

The increase in North American shale oil and gas production has created a glut of NGL/LPG.

The initial wave of new ethane crackers will absorb some of the increased ethane production. However LPG (propane and butane) is primary used as fuel product . The increased supply is outstripping the demand leaving prices depressed.

The production of NGL in North America is expected to continue increasing.

LPG Upgrading

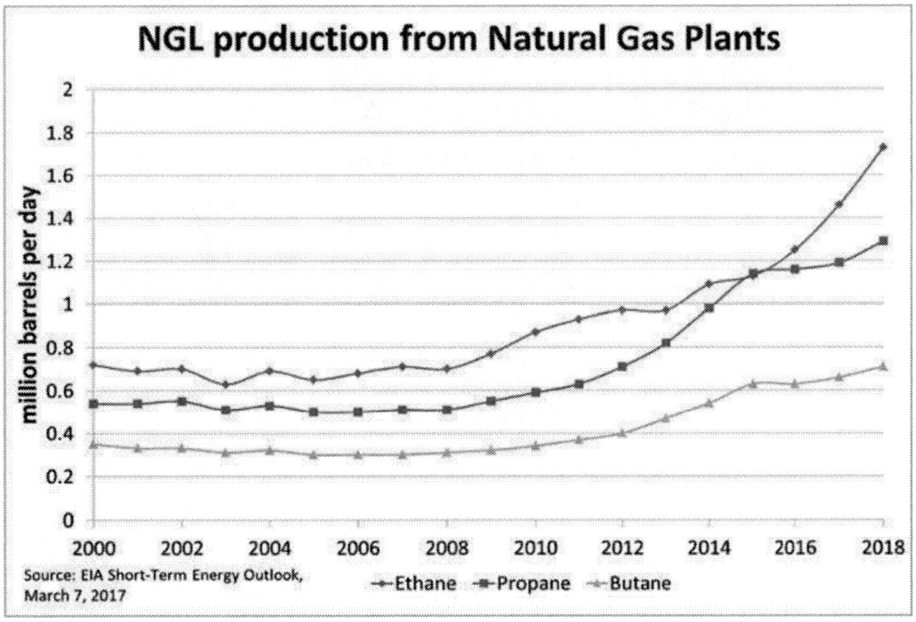

LPG Upgrading

The excess supply of LPG creates opportunities for use as a feed stock.

Propane dehydrogenation (PDH) and purification to polymer-grade propylene (PGP) is one process for upgrading propane that has received recent attention.

Polymer-grade propylene / polypropylene is a high value product. However, the process is capital and energy intensive.

Propane dehydrogenation (PDH)

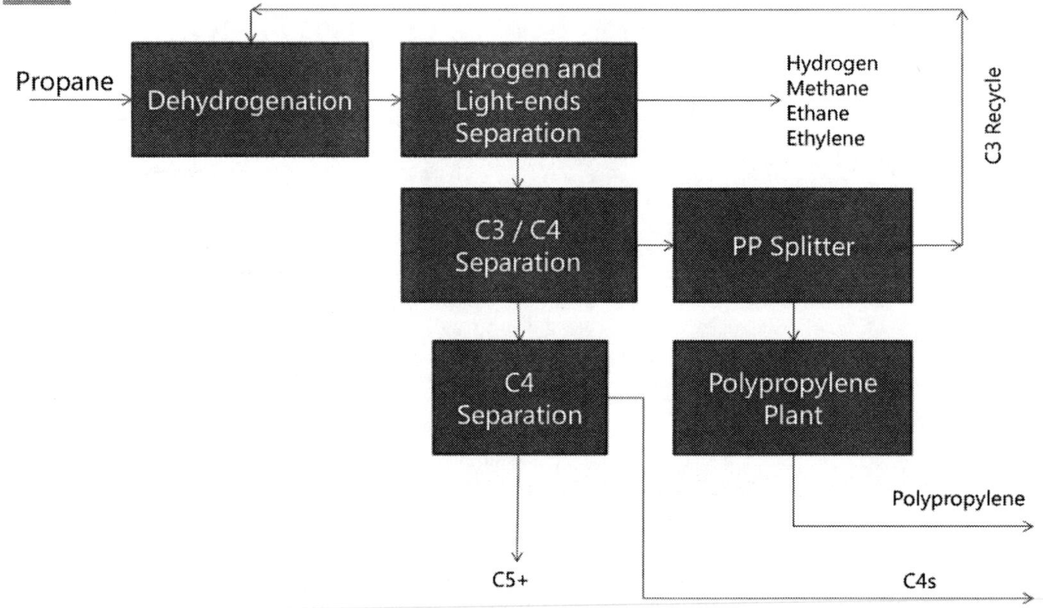

LPG Upgrading

Projects using this capital and energy intensive process are attractive when PGP and polypropylene prices are high enough . However, the overall market is limited and is unlikely to absorb the increases in propane and butane production.

Re-examination of existing technologies using a Process Intensification (PI) mindset leads to an alternate process for the creation of higher value products from propane and butane feeds stocks.

Propane and Butane to Liquids

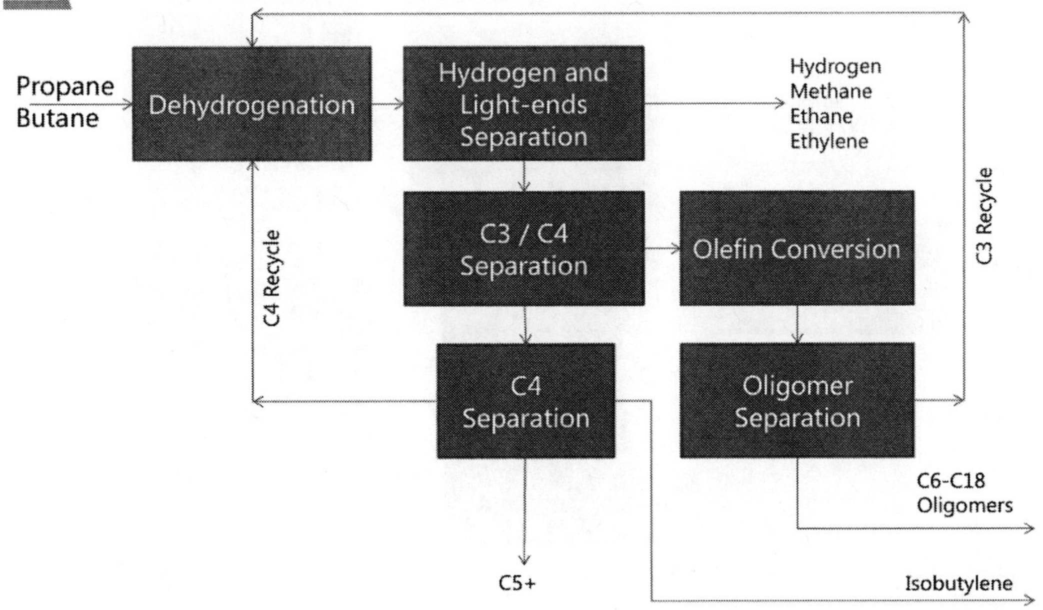

Propane and Butane to Liquids

In the alternate process olefin conversion is used to convert the C3 and C4 olefins into oligomers – molecules with a C6 to C18 carbon count or other lower volatility compounds.

This is the first PI step. The process for polypropylene production requires the distillation of the propylene to very high purity. This is very energy intensive because of the low relative volatility between propylene and propane.

Olefin conversion reactors can accept a mixed propane/propylene feed. The conversion of the propylene into lower volatility molecules makes the separation process much easier.

Propane and Butane to Liquids

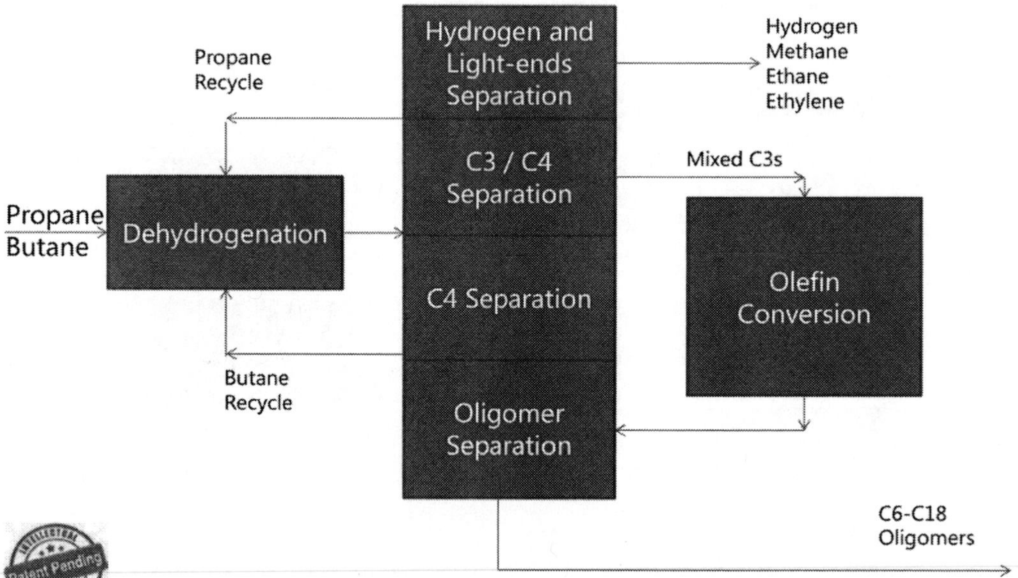

Propane and Butane to Liquids

The second PI step is to use DWC to integrate the distillation sequences and to configure the reactor to be an integrated side unit closely attached to the divided wall distillation column.

This configuration offers an attractive alternate for the utilization of excess propane and butane.

Patent Pending

Questions

????

More information will be presented in the Developments in C3-C5 Petrochemicals II

Converting LPG Into High-Value Products

Integrated process for dehydrogenation and conversion of propane and butane

Advisian

WorleyParsons Group

2017 AIChE® SPRING MEETING & 13TH GLOBAL CONGRESS ON PROCESS SAFETY

2017 AIChE® SPRING MEETING & 13TH GLOBAL CONGRESS ON PROCESS SAFETY

481018 Systematic Process Intensification Using Building Block Approach

Abstract:

Though research on process intensification has brought about many promising technologies including reactive distillation, divided wall columns, membrane reactors and task integrated columns, there is a lack of systematic methods for the identification and optimization of intensification alternatives at the flowsheet and equipment levels. Current superstructure-based process synthesis relies on pre-specified configurations and is unable to automatically construct novel equipment configurations for intensification. In this work, we depart from the classical unit operation-based representation of process units, flowsheets and superstructures and propose a new representation using fundamental building blocks which leads to a systematic method for simultaneous process design, synthesis and intensification. These building blocks can be associated with different process phenomena. An assembly of blocks of the same type obtains a classical unit, while an assembly of blocks with different types results in an intensified unit. This allows to systematically identify and incorporate many intensification pathways using a general block-based superstructure at the equipment and flowsheet levels. We obtain the intensified process by optimizing a performance metric for given raw materials and product specifications, material properties and bounds on flow rates. The overall intensification problem is formulated using a single mixed-integer nonlinear optimization (MINLP) model which is solved using commercial solvers. In this presentation, we will demonstrate the applicability of our approach using several process design and intensification case studies.

References:

[1] Demirel, S.E., Li, J., Hasan, M.M.F. Systematic Process Intensification using Building Blocks. Submitted to *Computers & Chemical Engineering*.

High performance steam super-heater in ammonia production lines
ID# 479284

Dr. Jörg Weidenfeller, Klaus Rothenpieler, Karsten Stückrath
SCHMIDTSCHE SCHACK | ARVOS GmbH, Kassel, Germany

1. INTRODUCTION

The steam system in Ammonia plants is often called as the heart of the production plant, since steam at different pressure levels is used for the process itself, as well as for the driving power for the process compressors. The steam is produced by waste heat recovery from the process and additional natural gas fired auxiliary boilers. Depending on the plant situation, modern steam systems are designed either for zero steam export or using export steam to produce electricity or – in case of an Ammonia / Urea complex - to supply steam or electricity to the urea plant since this plant is a consumer for both. To improve the overall process efficiency it is obvious that the steam system is a big lever.

A major part of the steam generation by waste heat recovery happens downstream of the secondary reformer in Ammonia production. The typical waste heat boiler package consists of a horizontal process gas cooler (PGC) with a steam drum on top followed by a steam super-heater (SSH) to improve the overall process efficiency. The waste heat boiler package operates with natural circulation of the water / steam. A typical arrangement for two parallel lines downstream an Auto Thermal reformer is shown in Fig. 1.

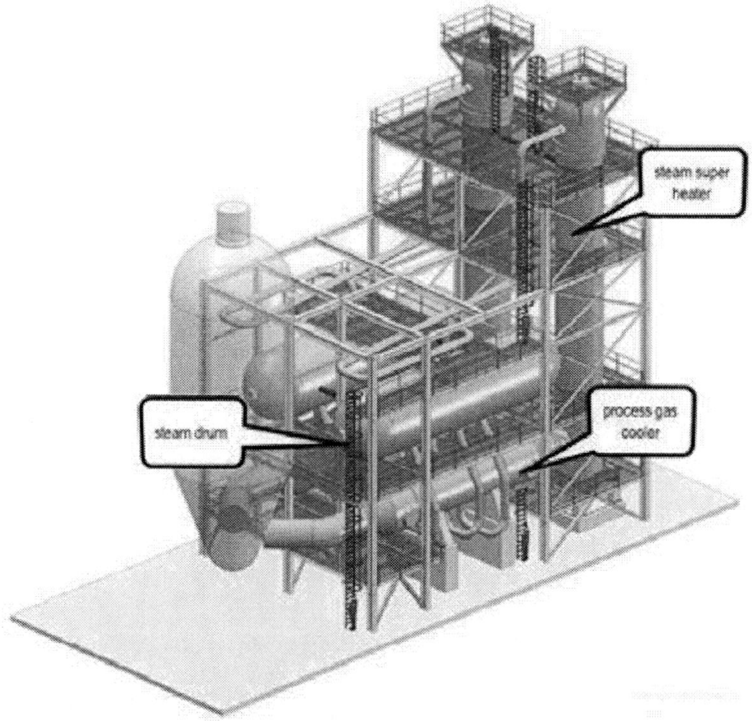

Fig. 1: SCHMIDTSCHE SCHACK | ARVOS solution for two parallel waste heat boiler packages

The process gas enters the waste heat boiler package typically at 800°C to 900°C and is cooled down to 350°C depending on further downstream process and requirements like the HT-shift. By recovering the heat from the process gas, high pressure steam is generated and super-heated. SCHMIDTSCHE SCHACK | ARVOS is solution provider for this kind of waste heat boiler package and designs and manufactures the complete package as well as single components. Based on executed and requested projects for all major licensors, the steam pressure and super-heated temperature is in a range of 100 bar to 130 bar and 330°C to 470°C, whereas the majority of those designs have high pressure steam at 125bar and 400°C, respectively.

1.1. Motivation for increasing the steam temperature

It is obvious that an increase in super-heated steam temperature has a beneficial impact on the overall process efficiency and cost effectiveness. Since SCHMIDTSCHE SCHACK | ARVOS is not in the role of a process designer, the further assessment only focusses onto the system of the waste heat boiler package by keeping constant the process conditions into the system. This is shown for the baseline or conventional scenario as well as for the improved scenario in the Heat-Temperature (Q-T) diagram in Fig. 2. It can be seen clearly, that the process gas inlet and outlet temperatures (as well as pressure and gas mass flow) are not changed for both scenarios, whereas the super-heated steam temperature is increased for the improved scenario. Since the total available heat by the process gas remains constant, and the required heat for the steam super-heating increases, consequently the remaining heat for steam generation in the PGC decreases.

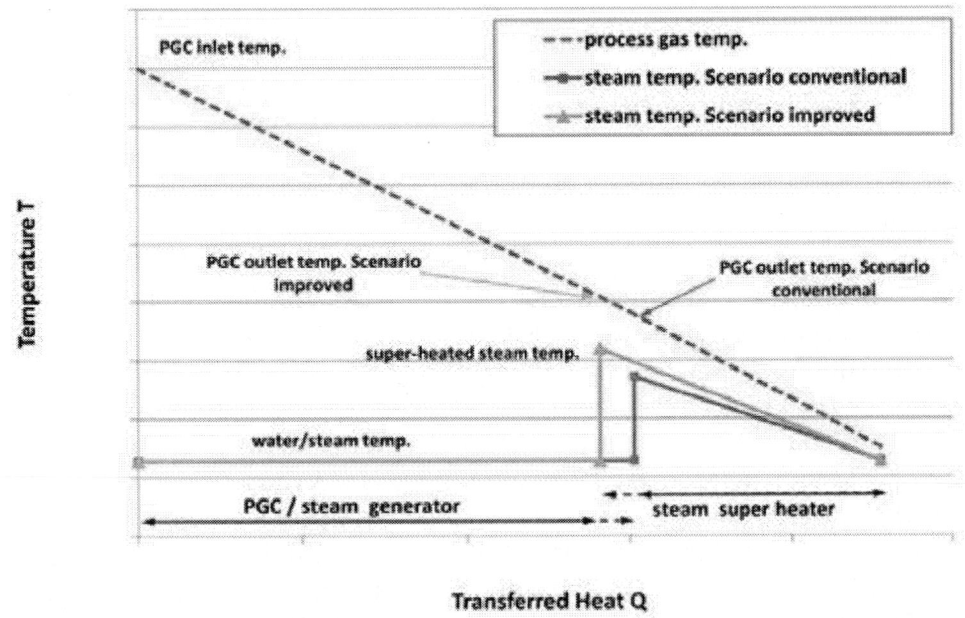

Fig. 2: Q-T diagram for waste heat boiler package, conventional and improved scenario

To quantify this effect, a baseline of a super-heated steam temperature of 400°C at 125bar is chosen. Increasing the steam temperature by 15% means an increase of the specific steam energy by 21% and consequently, as described above, a decrease in steam mass flow by 12% occurs. But overall the increase of the duty is 38% compared to the baseline scenario. For increasing the super-heated steam temperature by 30% the overall duty increases by 58%. Consequently this additional duty can be used for further electricity production or reducing the duty of natural gas fired auxiliary boilers.

A rough financial assessment for both improved scenarios is shown in Tab.1 and indicates clearly the financial benefit by increasing the super -heated steam temperature, either in saving natural gas or having additional revenue for electricity production.

But a vital requirement for those scenarios is a mechanical robust and reliable steam super-heater with high availability and long life time!

Table 1: Exemplary financial assessment of increasing super-heated steam temperature in SSH

Super-heated steam temperature	**Duty increase**	**Natural gas savings**	**Revenue for electricity production**	Ammonia process, 2200TPD plant, assessment for one year operation
400°C / 752°F	baseline	0	0	Assumptions: • Natural gas price 10,2 €/MWh (3,2 $/MMBTU) • Reimbursement for electricity 0.05 €/kWh • Plant operation: 7800 h/y • Efficiency steam to electricity 40% • Steam turbine expansion to 70°C
470°C / 878°F	7 MW	0.530 Mio US$	0.590 Mio US$	
520°C / 968°F	11 MW	0.900 Mio US$	0.810 Mio US$	

2. STEAM SUPER-HEATER, CONVENTIONAL DESIGN

The steam super-heater is the critical component of the waste heat boiler package. Therefore special care was taken for a robust and reliable design. As a result, the so called U-type steam super-heater was developed and established first in Ammonia plants by SCHMIDTSCHE SCHACK | ARVOS approx. 15 years ago and after that also in Methanol plants. Also, a limited number of other heat transfer equipment suppliers have applied a comparable principle.

Fig. 3 shows the design principle of the SCHMIDTSCHE SCHACK | ARVOS U-type SSH. The process gas enters into the gas inlet channel at the bottom and is guided upwards inside a central tube. At the central tube outlet near the tube sheet the gas is turned, enters the inner hot pass and flows downwards through the heating surface. At the bottom, the gas is turned again, enters the second cold pass of the U-type heating surface and flows to the top leaving the heating surface. After that the cooled gas flows downwards between the inner liner of the second pass and the outer pressure shell into the gas exit chamber. This means, that the hot process gas is at no point in direct contact with the pressure vessel wall, which is beneficial for the mechanical design of the outer shell.

The internal bypass valve controls the process gas outlet temperature by bypassing the hot gas directly into the gas exit chamber, where it mixes with the cooled gas to the required gas outlet temperature.

The super-heater operates in counter-current flow, which means that the saturated steam flows into the U-tube from the second to the first pass. The upper head is designed as a combined steam inlet and outlet chamber, where the super-heated steam is collected in an internal steam chamber.

Refractory is always a risk for cracks, material losses and damages require unexpected shut down due to immediate repair action. Therefore this design is focussed on a minimum use of refractory at the exit chamber only, which increases the availability of the steam super-heater and the Ammonia or Methanol production line, respectively.

The effectiveness of the heat transfer design is based on the counter current flow of the media. This has a beneficial impact onto the required size of the heating surface and therefore also onto the size, weight and cost of the steam super-heater.

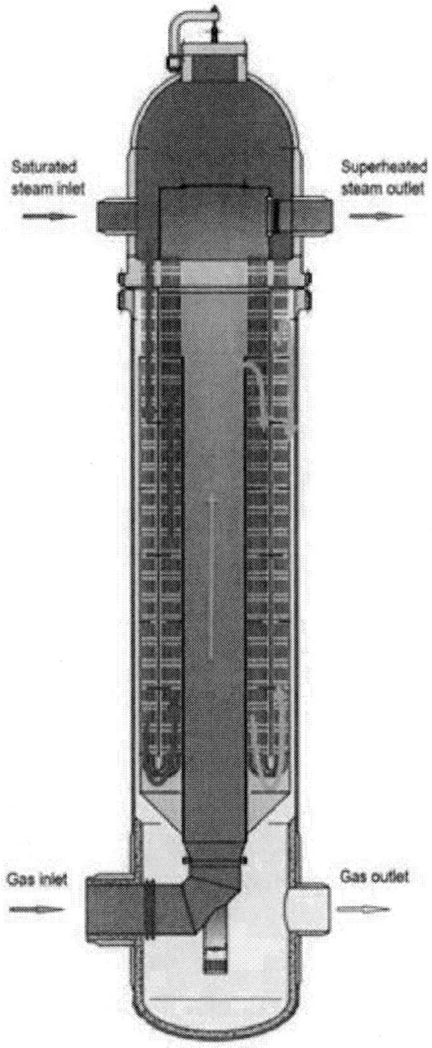

Fig. 3: Principle sketch of a SCHMIDTSCHE SCHACK | ARVOS U-type steam super-heater

Looking more in detail, the consequence of the counter current flow design is that both hot streams of the process media meet at one point, which is in this case close to the tube sheet at the first hot pass shown in Fig. 3. On the one hand, the hot process gas at shell side and the super-heated steam at tube side lead to high material temperatures for the U-tube, on the other hand, the high temperature at tube side is transferred into the tube sheet because of the hot U-tube leg. Only few possibilities are available to reduce this temperature impact by inserting ferrules, e.g., but this would increase also the pressure drop at steam side. Finally, the high temperature transfer into the tube sheet cannot be avoided.

This means that the temperature gradient caused by the media at shell and tube side is not only across the tube sheet, but a temperature gradient in the radial direction along the tube sheet is superimposed because of the U-tube principle with a cold and a hot leg.

The tube sheet, which is the heart of the steam super-heater, is loaded by the pressure of the steam and gas side as well as by the temperature gradients described above. A thick tube sheet is required to resist the primary stresses caused by the high steam pressure. But a thick tube sheet is also critical to thermal gradients, all the more to the thermal superposition mentioned above, which results in unfavourable secondary stresses.

For the current Ammonia applications and the latest process temperatures the conventional SCHMIDTSCHE SCHACK | ARVOS U-tube design is still feasible and reliable based on the long term experience for the thermal and mechanical design. But in case of improving the process efficiency by increasing the super-heated steam temperature beyond 470°, the conventional U-type design is no longer robust. The non-uniform temperature distribution mentioned above gets worse and consequently the tube sheet expands thermally non-uniform with additional thermal, secondary stresses.- This drives crack propagation at the tube sheet, tube to tube sheet welds and connecting welds. -The situation will get worse for cyclic operation, which is characterized by high temperature and load gradients.

3. HIGH PERFORMANCE STEAM SUPER-HEATER

SCHMIDTSCHE SCHACK | ARVOS has developed and manufactured a steam super-heater for a Methanation process downstream coal gasification, which has such high process parameters, that a U-type design is no longer reliable.

In general there are two design possibilities:

1. Cascade the steam super-heater in two exchangers, one hot and one cold apparatus
2. Choose one design, which fulfils reliability and life time criteria

Option 1 has the major disadvantage that an additional heat exchanger for the process is necessary and an additional failure risk is applied, which reduces the availability of the production line. Therefore SCHMIDTSCHE SCHACK | ARVOS decided to go for option 2.

The philosophy was not to develop a completely new design but to apply and combine proven design features to achieve a reliable solution.

The design for the pressure vessel of the above described U-type steam super-heater was used, since this design uses internal cooling features and a minimum of refractory. For the heating surface the U-tubes were not used but the so called bayonet design. This design was successfully applied by SCHMIDTSCHE SCHACK | ARVOS more than 20 years ago at comparable process temperatures as shown in Table 2.

Table 2: Combination of proven design features for the high performance steam super-heater

Process / Year	**Steam Side** [bar / °C]	**Gas Side** [bar / °C]	**Used design feature**
Ammonia / 2002, 2003, 2009	125 - 127/ 385 - 393	39 - 41 / 560	Pressure vessel of U-type
Styrol / 1994	0.9 / 840	3.1 / 642	Bayonet tube
Methanation / 1981	123 / 550	30 – 80 / 550	Bayonet tube
Methanation / 2014	130 / 520	30 / 620	Combination of design features to next generation SSH

The bayonet tube is a tube in tube concept and is therefore comparable with the SCHMIDTSCHE® double tube design, which is established in the Ethylene industry at more than 7000 transfer line exchangers for over 50 years. The bayonet tube originally was not invented by SCHMIDTSCHE SCHACK | ARVOS, but special thermal and mechanical design expertise and manufacturing experience is required to achieve a reliable solution.

The design was performed with three dimensional numerical flow and heat transfer simulations accompanied by mechanical structure analysis. In addition, experimental tests were performed to verify the boundary conditions for the simulation. A principal sketch of the bayonet type steam super-heater is shown in Fig 4.

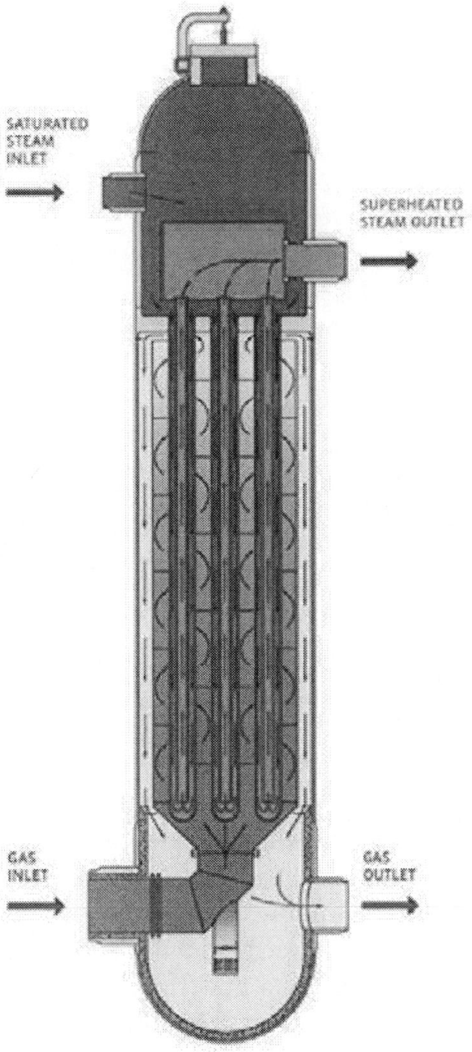

Fig. 4: Principle sketch of the SCHMIDTSCHE SCHACK | ARVOS bayonet type steam super-heater

The steam super-heater operates in counter current flow like the U-tube design. Saturated steam flows downwards in the annular gap between the outer and inner tube. At the bottom of the outer tube, the steam is already super-heated and flows again upwards in the inner tube. The process gas enters the heating surface from the bottom and flows upwards turned by disc and doughnut baffles, which optimize the heat transfer but has less pressure drop compared to standard baffle elements. At the end of the pass, the cooled process gas flows downwards between the inner liner and the outer pressure shell, which is comparable to the U-type design. An internal by-pass is also applied to control the gas outlet and steam outlet temperature, respectively. This by SCHMIDTSCHE SCHACK | ARVOS patented design has several thermal and mechanical advantages, which are described more in detail below.

The outer tube of the bayonet is connected to the tube sheet, which separates the steam and process gas side. This means that the tube sheet at steam side is only in contact with saturated steam, which is a major difference compared to the U-tube design. The super-heated steam flows in the inner tube of the bayonet, which is connected to the bottom of the steam outlet chamber. This design ensures that the outer tube as well as the inner tube can thermally expand independently without any constraint. Consequently no stresses due to thermal expansion occur at the outer and inner tube.

The major advantage of this design is demonstrated by the temperature distribution across the tube sheet in Fig 5 top (left side is grey scale, right side is colored scale).

Bayonet type Steam Side

Bayonet type Gas Side

U-type Steam Side

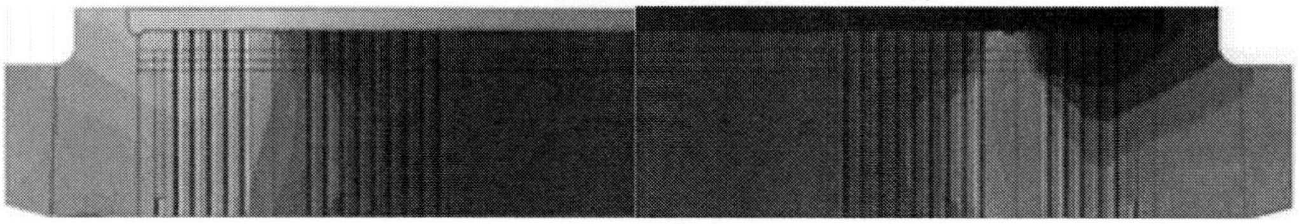

U-type Steam Side

Fig. 5: Temperature distribution across the tube sheet of the SCHMIDTSCHE SCHACK | ARVOS bayonet tube steam super-heater (top) and U-type (bottom). Same temperature scale light (blue) to dark (red) corresponds to cold to hot)

It can be seen clearly, that the tube sheet is cooled by saturated steam, whereas at the gas side the tube sheet is in contact with the already cooled gas. This leads to the very homogenous temperature distribution at low level. Consequently this enables a design for a small tube sheet thickness which has a positive effect on the thermal stresses in the tube sheet and the tube to tube sheet welds. In addition the smooth temperature distribution at low level has a beneficial contribution to the thermal stresses. For comparison only, the temperature distribution across the tube sheet of a U-tube design for the same process and steam temperatures is shown in Fig.5 bottom.

Both figures have the same temperature scale. **It has to be emphasized, that this U-tube tube sheet design is not robust and reliable but is only shown for comparison.**

In contrast to the top figure, this tube sheet is in contact with saturated as well as super-heated steam due to the U-tube concept. This leads, as described above, to a superimposed temperature gradient in the radial direction, which has negative impact on the thermal stresses in the tube sheet and tube to tube sheet welds.

3.1. Metal Dusting Aspects

In general, the risk for Metal Dusting is driven by the process gas composition, gas pressure and surface temperature of the metal, which is exposed to the process atmosphere. As discussed above, a higher super-heated steam temperature requires a higher gas inlet temperature into the steam super-heater and consequently the surface temperature of the heating surface increases. This increases the potential risk for Metal Dusting and the corrosion rate.

SCHMIDTSCHE SCHACK | ARVOS performed corrosion assessments for several material grades and temperatures for Ammonia applications. An example is shown in Fig. 6 for three different steam super-heating temperatures. The corrosion rates for the various materials are related to the corrosion rate of SA213-T22 for a metal temperature, which is reached for a steam super-heated temperature of 390°C / 734°F.

With increased metal temperature and super-heated steam temperature, respectively, the corrosion rate for T22 increases by a factor of 5 (470°C/878°F) and 16 (520°C/968°F), which is not acceptable. Therefore a material change to at least stainless steel is required to get moderate corrosion rates. To achieve for 470°C/878°F application the same corrosion rate as for 390°C / 734°F, the use of Nickel alloy is required. This is already state of the art. But for a further increase of super-heated steam temperature to 520°C/968°F and a corresponding increase in material temperature, the corrosion rate is nearly constant when Nickel alloy is applied for the heating surface.

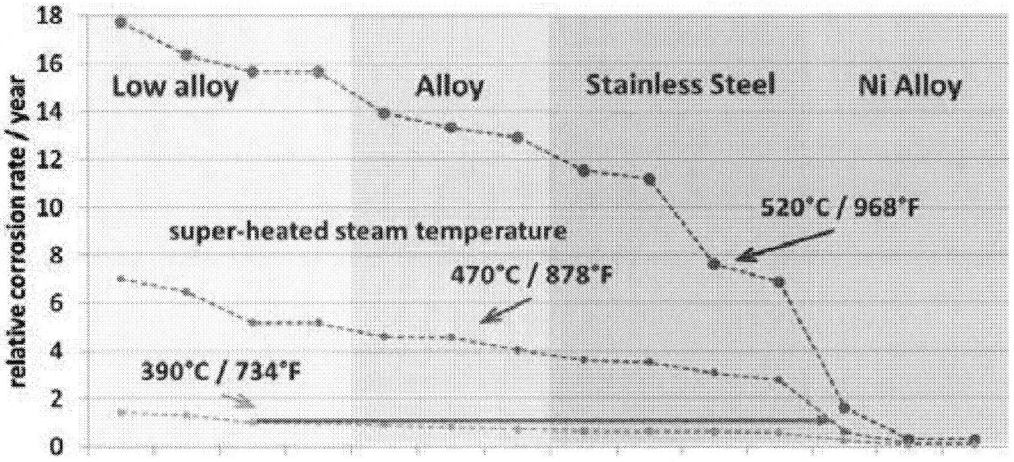

Fig. 6: Corrosion rates relative to SA213-T22 applied at steam temperature 390°C / 734°F

The shell and the tube sheet of the Bayonet type steam super-heater is only in contact with cold process gas as described above. With this inherent cooling feature, no risk of metal dusting for shell and tube sheet exists.

Therefore no further effort is required for this step towards high performance steam super-heating.

4. Conclusions

For the current process gas and steam conditions in Ammonia and Methanol applications, the steam super-heater with U-tube design is still robust and reliable. However, based on the process flow schema, the U-type is close to its mechanical threshold. For that reason, SCHMIDTSCHE SCHACK | ARVOS has developed steam super-heaters for the next generation of process temperatures, which are currently applied in a Methanation processes (see Fig. 7). The new apparatus combines proven design features to ensure an evolutionary and reliable step in steam super-heater development. Special care was taken in the process flow concept, the design of the heating surface and the tube sheet to achieve a mechanically robust design, which supports a long life time and reliability of the super-heater. The application of internal cooling features significantly reduces the use of refractory, which positively supports high availability of the apparatus.

Therefore this design fulfils the critical criteria: life time, reliability, and availability, which are important factors to achieve the planned return on investment during the operating time of the plant.

This new steam super-heater therefore provides the licensor and the customer a more reliable potential for improving the overall process efficiency.

Fig. 7: High performance steam super-heater ready for shipment

Heat exchanger design with high shear stress:
Evaluating the impact of Performance of Crude Pre-heat Trains

2017 Spring Meeting
San Antonio March 28, 2017

F. Coletti, E. Diaz-Bejarano and S. Macchietto
Hexxcell Ltd.

Crude Distillation Unit

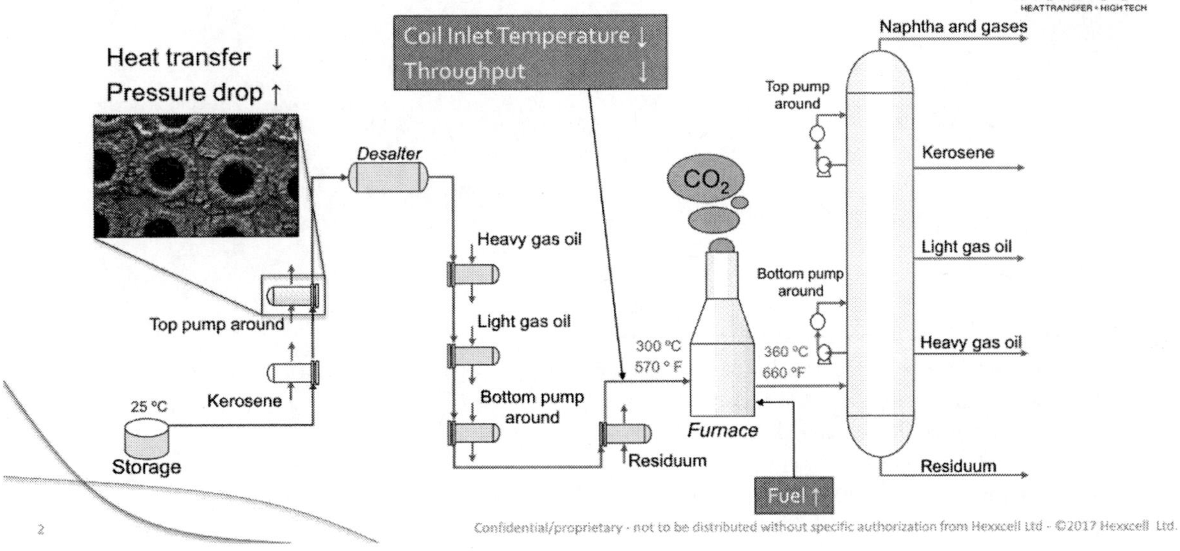

Complex, interacting phenomena

hexxcell
HEATTRANSFER • HIGH TECH

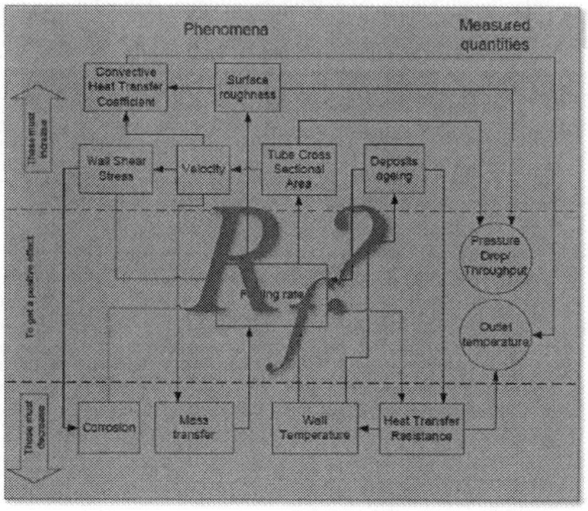

Coletti and Macchietto, I&ECR 2011

hexxcellSTUDIO™ – Under the bonnet

hexxcell
HEATTRANSFER • HIGH TECH

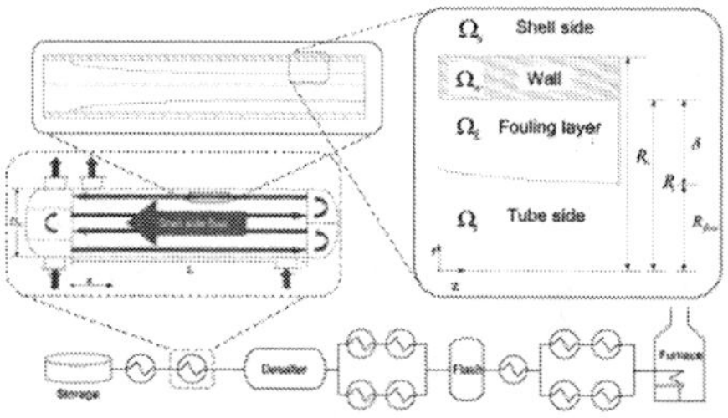

$$\frac{\partial}{\partial t}(A_{p,c}\rho c_p T) = -\frac{\partial}{\partial z}(A_{p,c} u \rho c_p T) + \frac{\partial}{\partial z}\left[A_{p,c} k \frac{\partial T}{\partial z}\right] + P h_i (T_w - T)$$

$$\frac{\rho_s c_{p,s}}{k_s}\frac{\partial T_{s,e}}{\partial t} = \frac{1}{r}\frac{\partial T_{s,e}}{\partial r} + \frac{\partial^2 T_{s,e}}{\partial r^2}$$

$$\frac{\partial}{\partial t}(\rho_s c_{p,s} T_s) = \frac{\partial}{\partial z}(\rho_s c_{p,s} T_s u_s) + \frac{\partial}{\partial z}\left(k_s \frac{\partial T_s}{\partial z}\right) + \frac{1}{A_s}\sum P_s h_s (T_s - T_w)$$

$$\frac{dR_{f,e}}{dt} = \alpha \mathrm{Re}^{-0.66} \mathrm{Pr}^{-0.33} \exp\left(\frac{-E}{RT_{f,e}}\right) - \gamma\tau$$

$$\frac{d\delta(z,t)}{dt} = \lambda_f \frac{dR_f(z,t)}{dt} \qquad \frac{dP(z,t)}{dz} = C_f(z,t)\frac{\rho}{4}\frac{v^2(z,t)}{R_{flow}(\delta)}$$

Coletti and Macchietto, I&ECR 2011; Diaz-Bejarano et al. AIChE J. 2016

<u>Main features:</u>

- Truly *multiscale*, *dynamic* and *incremental* model (distributed in space)
- Simultaneous *thermal* and *hydraulic* impact of fouling
- *Prediction* of fouling behaviour as a function of process conditions
- Tube-side and shell-side fouling and ageing
- Use of primary measurements (i.e. T, F, ΔP) not derived ones (e.g. R_f)
- Technical and *economic* results
- Simulation of *mechanical* and *chemical* cleaning

Approach

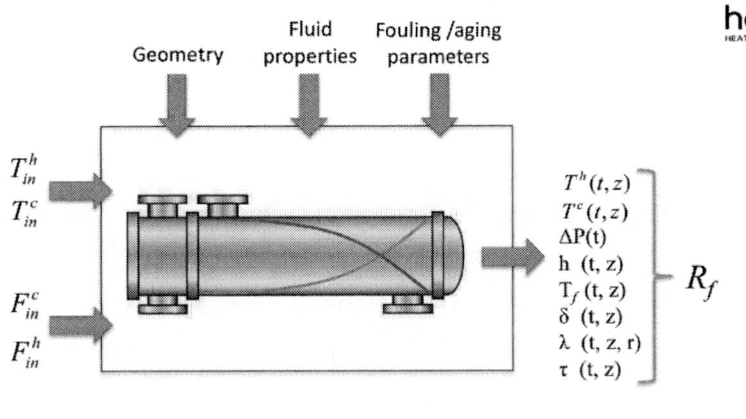

Focus on primary quantities (T) NOT on derived ones (R_f)

Case study – ExxonMobil refinery

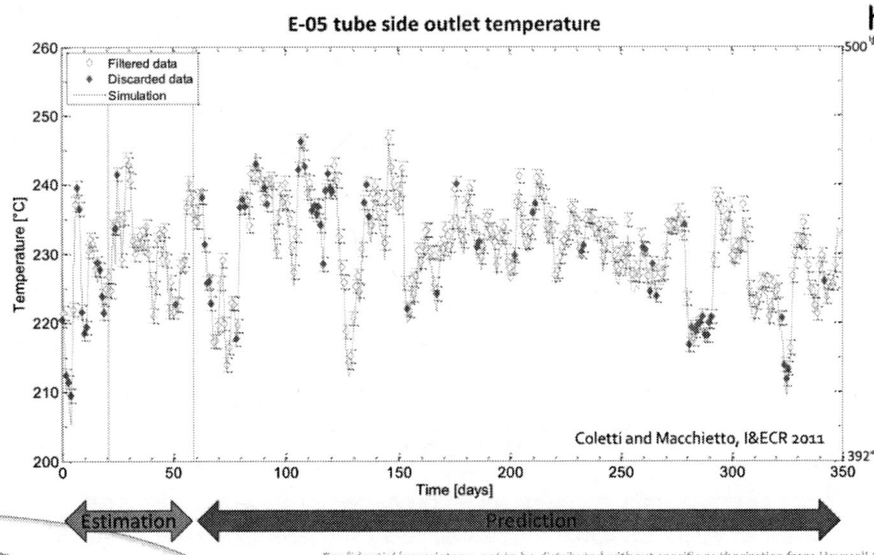

hexxcell
HEATTRANSFER • HIGHTECH

Heat exchanger design with high shear stress: reducing fouling or throughput?

Heat exchanger retrofit

hexxcell
HEATTRANSFER • HIGHTECH

- Increasing shear stress decreases fouling:

$$\frac{dR_f}{dt} = \alpha \,\mathrm{Re}^{-\beta} \exp\left(\frac{-E}{RT_f}\right) - \gamma\tau$$

Ebert and Panchal (1995)

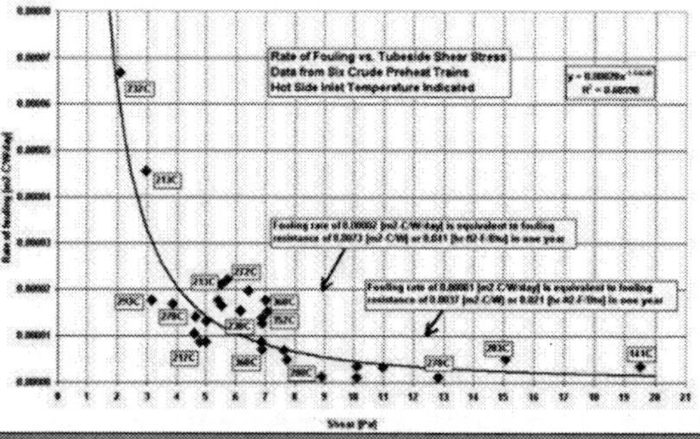

Joshi *et al.* (2009)

Increase in shear stress = increase in ΔP → production loss?

HEX retrofit case study

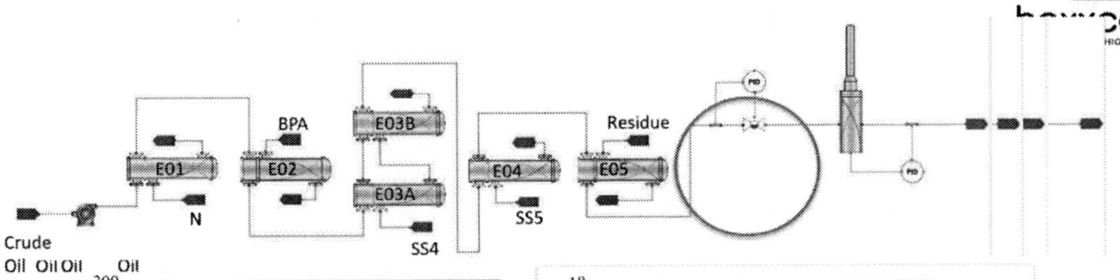

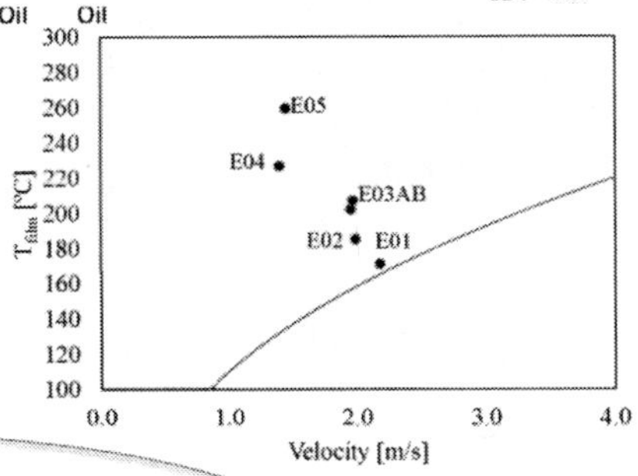

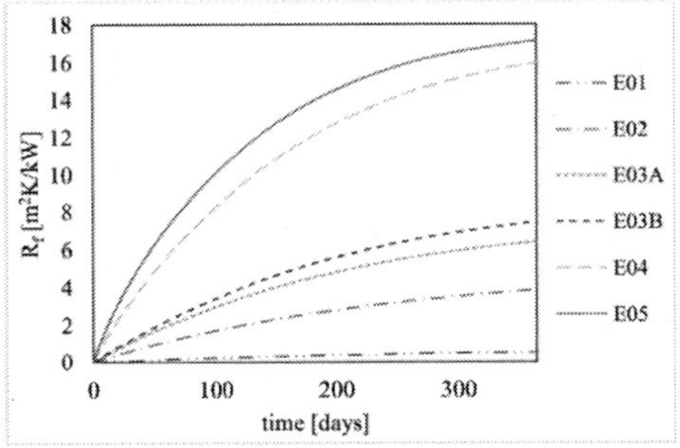

HEX retrofit case study

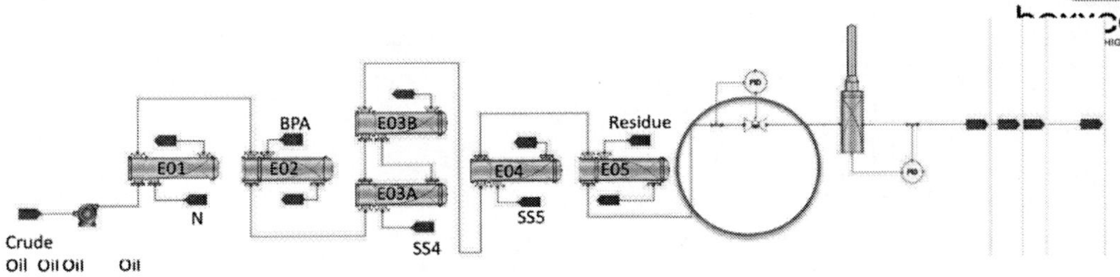

Effect of retrofitting E05 to increase shear stress:

I. Increase number of pass
II. Reduce tube diameter

Threshold fouling plot

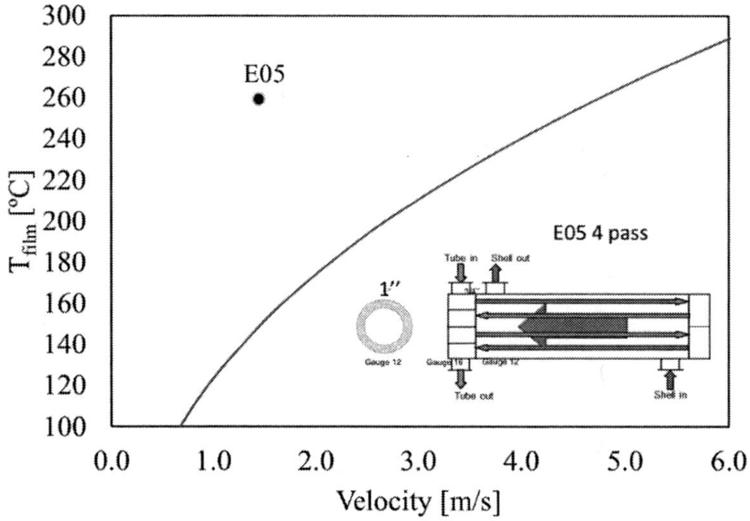

11 Confidential/proprietary - not to be distributed without specific authorization from Hexxcell Ltd - ©2017 Hexxcell Ltd.

Smaller tube diameter

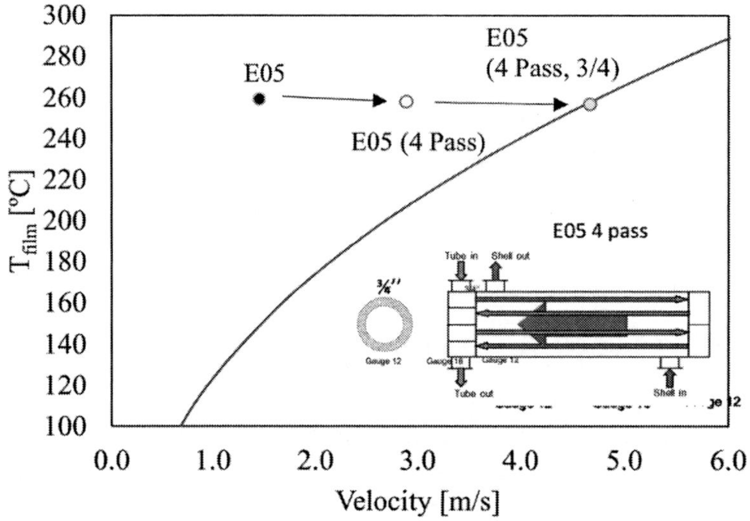

12 Confidential/proprietary - not to be distributed without specific authorization from Hexxcell Ltd - ©2017 Hexxcell Ltd.

HEX retrofit case study

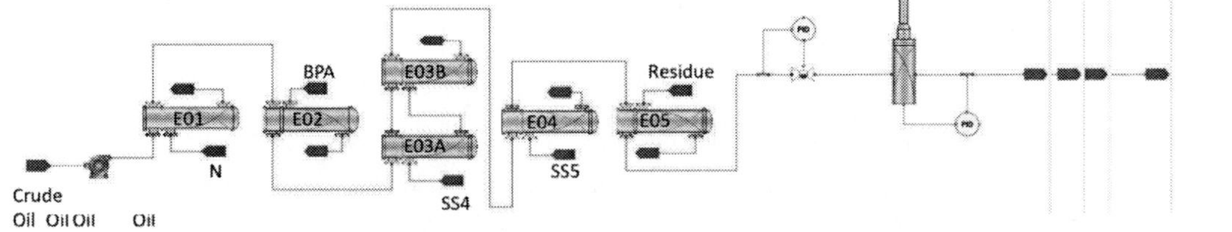

- Effect of retrofitting E05 to increase shear stress:
  - I. Increase number of pass
  - II. Reduce tube diameter
- Varying hydraulic resistance of the network:

Case 1 < Case 2 < Case 3

low medium high

Case study 1: low hydraulic resistance

Pressure drops

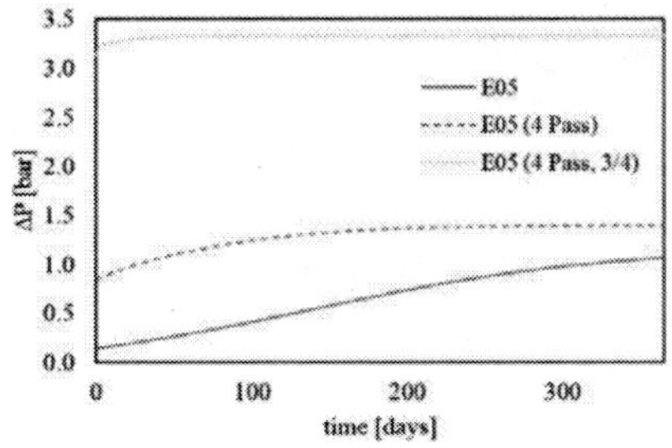

Valve opening

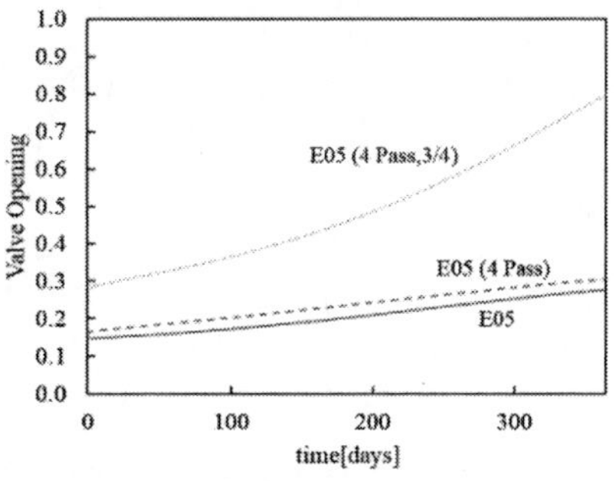

Hydraulic limit is not reached

Case study 1: low hydraulic resistance

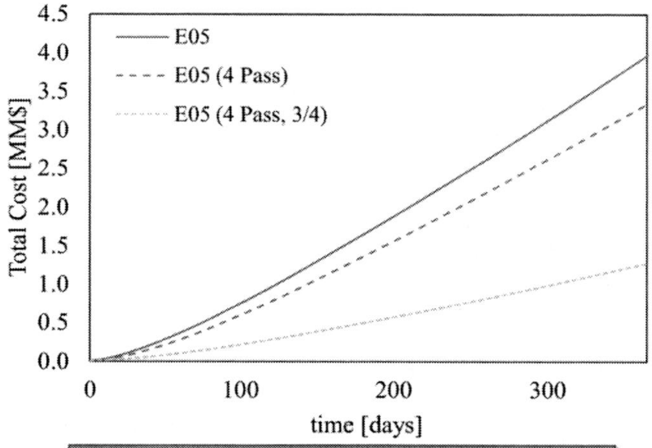

Costs only associated with extra fuel

HEX retrofit case study

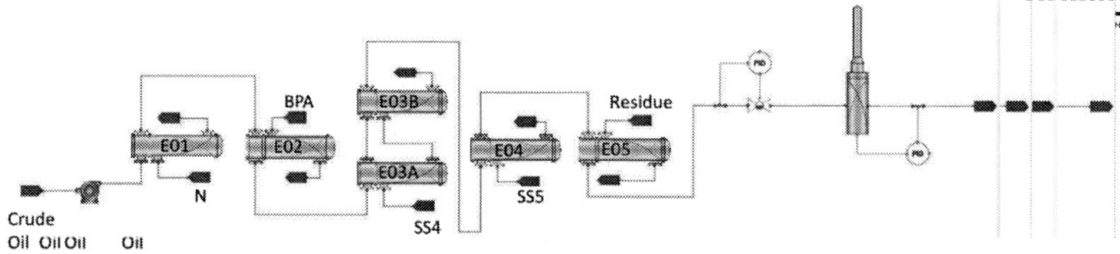

- Effect of retrofitting E05 to increase shear stress:
  I. Increase number of pass
  II. Reduce tube diameter
- Varying hydraulic resistance of the network:

Case 1 < Case 2 < Case 3

low medium high

Case study 2: medium hydraulic resistance

hexxcell
HEATTRANSFER • HIGH TECH

Pressure drops

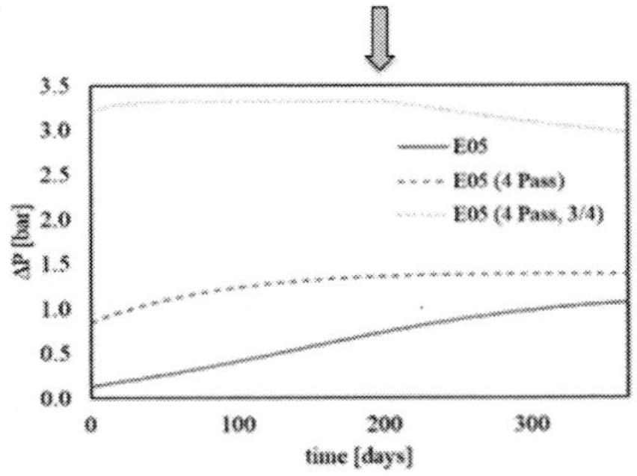

Valve opening

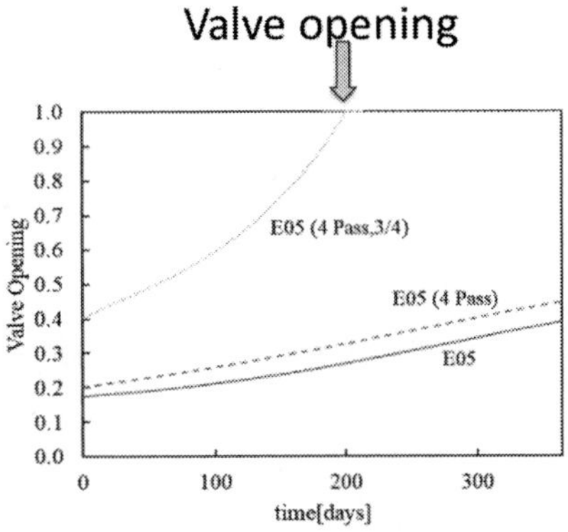

Hydraulic limit reached only with 4 pass ¾' tubes

Case study 2: medium hydraulic resistance

hexxcell
HEATTRANSFER • HIGH TECH

Throughput

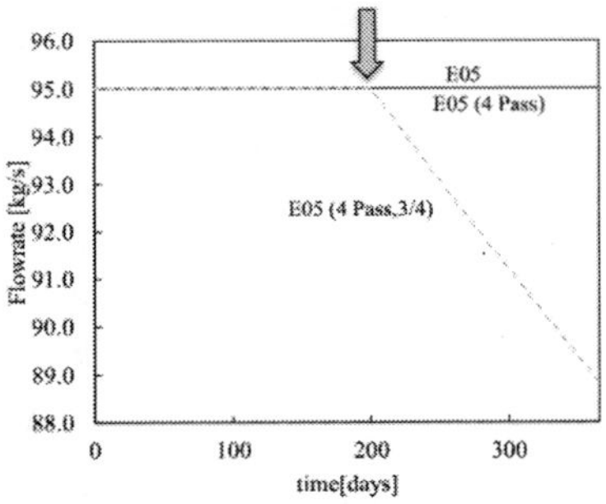

Total costs

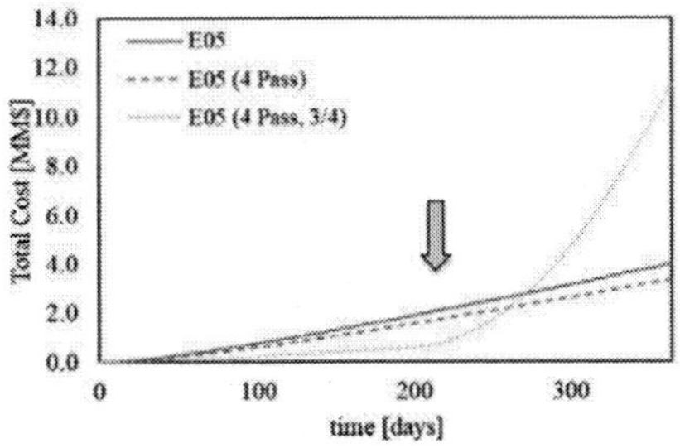

4 pass ok - 4 pass & ¾'' tubes hits hydraulic limit

HEX retrofit case study

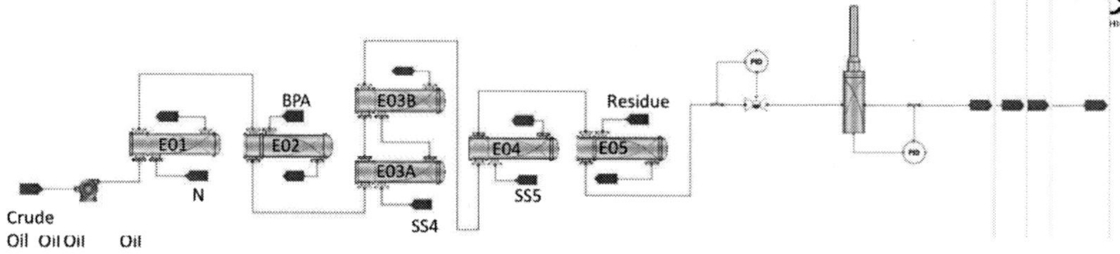

- Effect of retrofitting E05 to increase shear stress:
  I. Increase number of pass
  II. Reduce tube diameter
- Varying hydraulic resistance of the network:

Case 1 < Case 2 < Case 3
low medium high

Case study 3: high hydraulic resistance

Pressure drops

Valve opening

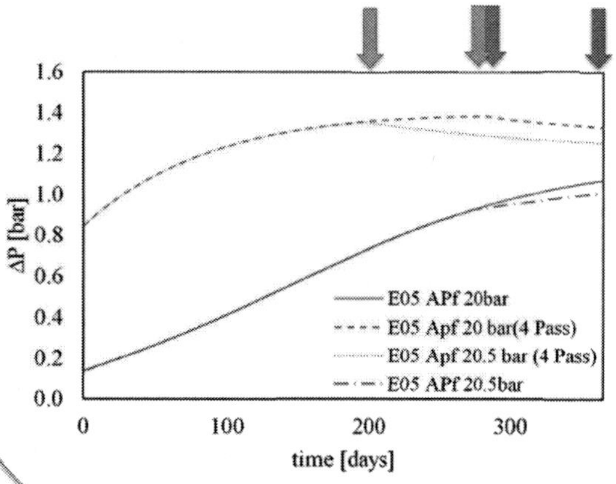

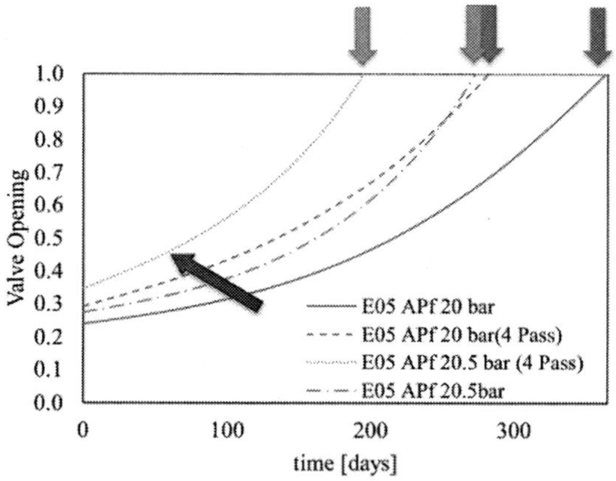

Case study 3: high hydraulic resistance

hexxcell
HEATTRANSFER • HIGHTECH

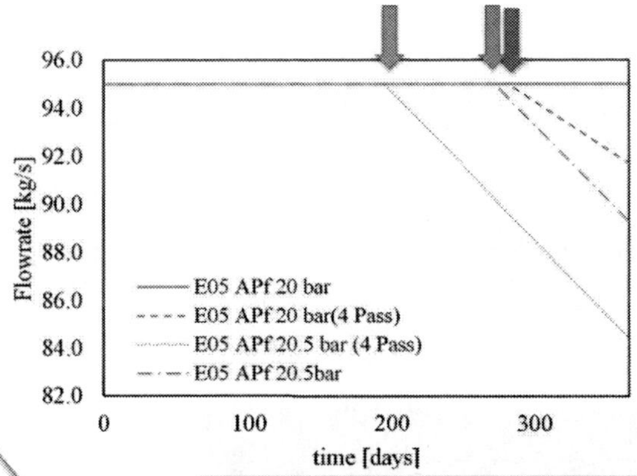

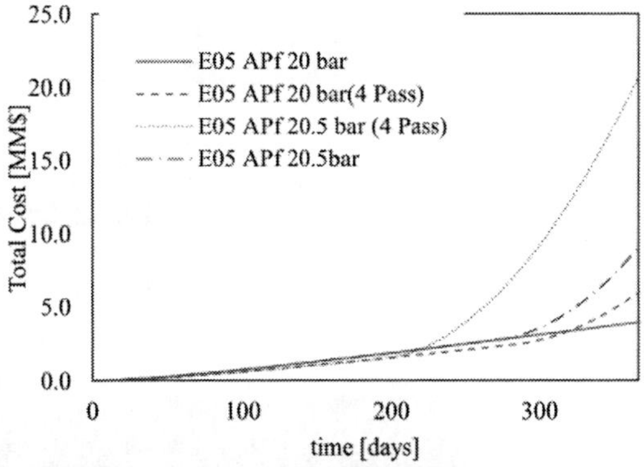

Hydraulic limit reached with any high shear retrofit

Using plant data - Dynamic Retrofit Test™

hexxcell
HEATTRANSFER • HIGHTECH

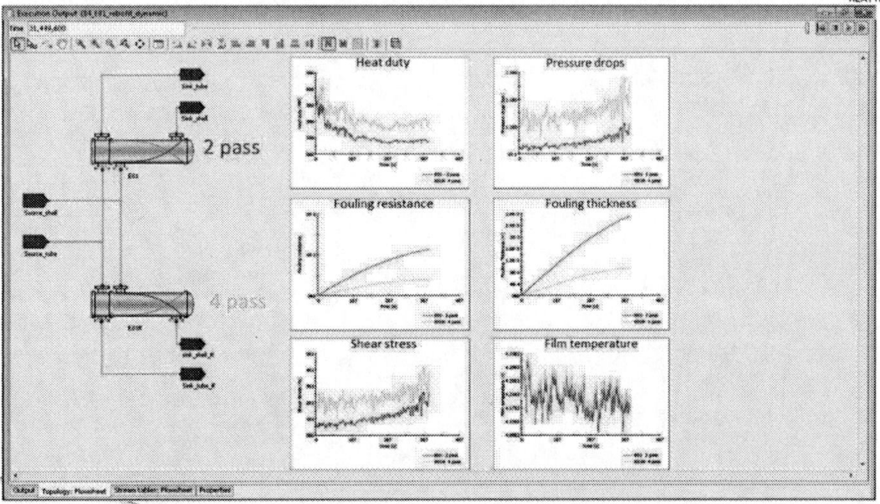

Conclusions

hexxcell
HEATTRANSFER • HIGHTECH

Heat exchanger design with high shear stress: reducing fouling or throughput?

It depends!

Trade-offs need to be assessed to avoid costly consequences

23 Confidential/proprietary - not to be distributed without specific authorization from Hexxcell Ltd - ©2017 Hexxcell Ltd.

Thank you!

f.coletti@hexxcell.com

hexxcell
HEATTRANSFER • HIGHTECH

479928 Performance of the Artificial Neural Networks for Estimation of Pool Boiling Heat Transfer Coefficient of Alumina Water-Based Nanofluids

Abstract:

Due to the transformation of the latent heat during the boiling process, it appeared as an efficient process for heat transfer purposes. Boiling heat transfer coefficient is the most important parameters for calculation of the amount of energy that can be transferred by the boiling process. Recently, new class of fluids namely nanofluid is widely used to improve performance of the boiling process. In spite of the large amount of experimental researches on the pool boiling heat transfer coefficient, theoretical and modeling basics of this parameter was less taken into account.Therefore, in this study, pool boiling heat transfer coefficient of water-alumina nanofluid is tried to be predicted using artificial neural networks. Excess temperature, operating pressure, diameter and weight fraction of nanoparticles are used for modeling of the considered coefficient of the nanofluids. At first, four different types of the artificial neural networks, and various empirical correlations are employed for estimation of the pool boiling heat transfer coefficient of water- alumina nanofluid. Thereafter, the best intelligent model is selected through comparing the predictive accuracy of various intelligent/empirical approaches.

The results show that a two-layer feedforward neural network with fourteen hidden neurons is the best model for estimation the pool boiling heat transfer coefficient of water-alumina nanofluid. This model was able to predict the considered coefficient with overall MSE=3.98, AARD%=9.80, and R^2=0.98.

480098 Advanced Process Control in a Mobile Pilot Plant for the Evaluation of Long-Term Dynamic Performance of a CO_2 Capture Process Under Industrial Conditions

Abstract:

The removal of carbon dioxide (CO_2) from industrial gases is of great interest and is an important technology for the reduction of greenhouse gas emissions. One common method is the application of reactive solutions in absorption/desorption processes. This technology is well-established and has commercially been implemented in industry [1]. However, most processes suffer from unstable operation conditions, such as fluctuations in feed gas compositions or changes in the required load [2]. In addition, non-optimal operation conditions for regeneration lead to increased costs for the whole process. The flexible operation of plants for carbon capture is of great interest and is essential for future applications. The investigation of the dynamic performance of CO_2capture processes by dynamic test campaigns provides information to compensate changes in the operation conditions. For this purpose two scenarios are considered. Firstly, fluctuations in the gas feed, such as changes in the flow rate or gas compositions, are observed. Due to the coupling of the process to power plants, daily changes of the load in a power plant require the flexible operation of the carbon capture process.

Therefore, a modular and transportable pilot plant is designed and constructed by the *Process Dynamics and Operations Group at Technische Universität Berlin*[3]. Regarding varying compositions of the gases and the operation of the pilot plant a flexible and reliable design of the process is required. In addition to that, the plant is operated under real conditions at a steel mill in Duisburg, which highlights the importance of a flexible implementation.

These aspects need to be considered for the conceptual design, the construction of the pilot plant, and the process control design. For this reason, the concept of the pilot plant contains two absorption columns with a total height of 3 m of structured packing and a diameter of 0.11 m each. The desorption column for the regeneration contains 2.5 m packing with the same diameter. Preceding pretreatment columns are installed to maintain stable operation conditions at the inlet of the absorber and the removal of acidic components. Following the absorption and desorption columns posterior treatment columns are implemented to ensure sustainable process conditions at the gas outlet and to reduce the loss of the amine solution. The pretreatment and posterior treatment columns contain 1 m of structured packing each. The pilot plant is automated using ABB´s Freelance 700F process control system. To ensure reliable operation conditions and to provide the required information for the process control more than 30 temperature, 13 pressure, 8 level, and 13 flow indicators are installed.

For the studies of the dynamic responses of the carbon capture process advanced process control loops are applied by the implementation of soft sensors. Thus, the CO_2 removal is controlled by the feed flow rate of the amine solution to the absorption. Another issue is the change in the gas load. Due to the coupling of the process to power plants, the operation of the CO_2capture process is adjusted to the load of the power plant. For this reason, a switch in operation modes is considered, which compensates partial gas loads and leads to reduced heating duty for the desorption.

The main purpose of this contribution is the systematic evaluation of dynamic responses of the pilot plant towards unstable operation conditions, such as fluctuations in the gas flow or changings gas loads. Therefore, advanced process control loops are introduced by the implementation of soft sensors. These aspects are considered and implemented in a pilot plant.

[1] S. Ziaii, T. F: Edgas, G. Rochelle „Dynamic Modeling to Minimize Energy Use for CO_2Capture in Power Plants by Aqueous Monoethanolamine", Ind. Eng. Chem. Res. 2009, Vol. 48, 6105-6111

[2] P. Tait, B. Buschle, I. Ausner, P. Valluri, M. Wehrli, M. Lucquiaud "A pilot-scale study of dynamic response scenarios for the flexible operation of post-combustion CO_2capture", International Journal of Greenhouse Gas Control, Vol. 48, 2016, 216-233

[3] M. Müller, G. Wozny, H. Thielert „Innovative Produkt- und Prozessentwicklung mittels mobiler und modularer Mini-plant Technik", Jahrestreffen der Fachgemeinschaft Prozess-, Apparate- und Anlagentechnik, Fulda 2011

Optimal Design of Separations for Pilot Plants and Laboratory Bench Units

Joseph B. Powell, PhD, Shell Chief Scientist – Chemical Engineering, Houston, TX

Abstract

Separations for pilot plants and laboratory bench units typically will not scale directly to commercial. Attempts to reproduce the actual extent of separation targeted for commercial operation may lead to unnecessarily large and expensive pilot or demonstration units. It is important to design research and development units which can provide target recycle compositions to test integrated process chemistry, and to provide separation data for use in design of commercial units. This seminar will present some examples of pilot and bench units used in design of commercial separations via distillation, liquid extraction, adsorption and ion exchange, and crystallization, with emphasis on use of shortcut methods augmented by numerical modeling.

Introduction

Pilot plant and demonstration projects are expensive relative to bench-scale experiments and modeling (Figure 1). Time required to build, safeguard and operate larger units can also be detrimental to project economics, where R&D expenditure prior to commercial production results in negative cash flows which offset project value. In some cases, cost and timing for demonstration plants completely offset the future value of the project, and become a “show stopper” for the technology development. Optimization of process development and costs requires careful consideration of what needs to be piloted and why?

Technical Risk Registry

Assessment of technical risks involved in process development for a new concept can help prioritize R&D activities, and assess scales needed for de-risking. Often new feed stocks, or catalysts or reactor concepts comprise the principal challenges and risks to address. Primary separations may have critical new technology development needs for some concepts (reactive distillation, selective crystallization), but may be noncritical for others especially where they can be calculated from physical property data using process simulators. Often however, it is the trace component or secondary separation problem that impacts process viability. “Trace chemistry” often is not apparent until one runs an “integrated pilot plant” to discern problems in separation, which may appear as onset of foaming, or failure to achieve target purity.

Proper sizing of pilot plants and possible demonstration units can thus depend on:

1. Scale required to de-risk hydrodynamics or residence time distribution

2. Requirements for running an integrated process to determine impacts of trace chemistry and build-up to steady state compositions.
3. Requirements for producing product for product or market development.

For product development (#3), larger scale pilots or demonstration plants may be needed. For testing and scale-up of process steps (#1 and #2), however, it may be necessary to test a specific unit operation at scale (fluidized bed reactor), but the primary separation steps may be calculable from first principles and not require piloting in the same type of separation as proposed for commercial, or at the scale needed to match the rest of the pilot design at steay state.

Distillation Optimization

Distillation is the most common method for separations in the process industries, and performance can generally be calculated *a priori* if components are known, using process simulators with known physical properties. Trace chemistry remains a challenge for process development. If bulk compositions are known, then process simulations can be used to predict temperature profiles for small pilot plant columns. One can then conduct pilot studies in a relatively small column (e.g. 25 mm diameter) using a calculated temperature profile imposed by direct wall heating. If the compositions are not known, then a nearly adiabatic vacuum-jacketed glass column can be considered at larger diameter (75-mm) to avoid significant heat losses. This increases the required throughput nearly 10-fold, or requires operation at higher reflux ratios.

For one process development program, a 7-meter tall x 75-mm distillation column was constructed for a new product development. Use of protruded metal packing provided an estimated 80 theoretical stages for separation. A smaller once-through reactor / extractor pilot was operated to produce intermediate product for purification by distillation. Accumulation of this product from the smaller pilot allowed inventory of the larger-sized distillation column for operation in batch reflux mode (infinite recycle ratio). Addition of known markers and use of shortcut distillation theory [1] allowed fitting of thermodynamic property parameters for the unknown components, and could establish whether desired purity product could be produced as a side draw.

For this case, on-spec product was not produced. A smaller batch distillation unit (500-ml, 250 mm column height) was streamed, and a batch distillation sequence was conducted. It was found that the concentration of a volatile component was increasing with each successive distillation cut, and not separated from the heavier key component to be purified (Figure 2). This indicated that undesired reactive distillation was occurring, whereby heavy components were reacting in the column to produce light byproducts. An inverted distillation sequences was thereby designed and tested in the batch still, where the heavy bottoms were first topped to produce a majority distillate, and the tops distillate then redistilled to produce product where purity now increased with increasing extent of distillation.

Given that the batch column afforded only about 10 theoretical stages, the pilot column was used to distill the larger accumulated inventory of material, first in topping mode, with distillate then recycled to conduct a second distillation free of the initially removed heavy ends. On spec product was now demonstrated, using the information learned in the small batch distillation sequence.

In this manner, the separations studies needed for process development were conducted using a reaction system front end that was only $1/10^{th}$ the size of what would have been required if a fully continuous, nearly adiabatic distillation train were required to obtain integrated performance data. A larger distillation column was constructed, but this did not drive the size of the continuous reactor pilot unit, as it was not directly coupled to the reactor / liquid extraction front end. Rather, integrated or accumulated product from the continuous reactor pilot provided batch feed to the distillation pilot, stand-alone unit. Two distillation columns were found to be needed for the continuous process to conduct the inverted distillation sequence needed to overcome undesired reactive distillation. By running distillation in semi-batch mode, construction of only one column was needed, with collection of topped intermediate product from an initial distillation to send through the same column to demonstrate performance of the second, finishing column. Process development costs and timing requirements were both substantially reduced.

Entrainment, Skin Temperature, and Purge rates

Process equipment often does not scale up because of inability to keep linear velocities and residence times constant during scale up. Thus, a process separator for disengaging gas-from-liquid or two liquids following extraction with a rule-of-thumb 10-minutes residence time, will give a much smaller linear velocity for a pilot unit with overall equipment sizes well less than a meter, vs. commercial units with a characteristic dimension of several meters. While coalescence and separation is designed on the basis of residence time, phase entrainment or onset of foaming is a function If linear velocity. Thus, commercial equipment will often foam or entrain, where pilot or lab bench equipment did not, due to higher linear velocities required to achieve the same residence times.

One can calculate this offset however and test for entrainment or foaming using semi-continuous testing with representative intermediate product, where recycles are at least partially closed. Thus, the pilot or lab equipment can be operated for short periods of time at high linear velocities to test for onset of foaming or entrainment, without having this feature dictate the scale of the overall pilot program.

Similar issues hold true for skin temperatures for heat exchangers or reboilers. Lab and pilot equipment readily provide high surface area per unit volume for heat transfer. Larger scale commercial equipment does not provide similar area, on the basis of a surface A to volume V ratio of $[A \propto d^2]/[V \propto d^3] = 1/d$. One can however conduct specific experiments e.g. in stirred autoclaves with fixed wall temperature to examine the impact of calculated skin temperatures for commercial designs, on decomposition kinetics. Kinetic measurements are thus used to calibrate a model for prediction of impact on the specific commercial design assumption for heat transfer area.

Samples taken from pilot plants or bench reactors can represent huge purge rates for impurities relative to commercially practiced processes. One can examine the buildup of impurities and compare with calculated concentrations basis purge rate differences upon scale up, and then either concentrate or dope the corresponding impurities to consider impacts on foaming or emulsions, or ability to separate in purification. Commercial performance can again be calibrated and modeled, despite the implausibility of operating the pilot unit at low commercial purge rates.

Using known performance equipment for pilot or bench studies

If for example a packed liquid extractor or distillation column is used for separation under pilot or lab conditions, the number of theoretical stages may not be known, unless calibrated via known markers as shown for the previous distillation example. However, it may not be necessary to pilot the exact type of separator to be deployed commercially, if commercial performance can be calculated once appropriate property or separation factor data is available. Thus, use of one or more mixer-settlers to rigorously define the number of stages and measure liquid-liquid extraction equilibria may be more valuable at the pilot stage, than using a packed or sieve-tray extractor that would attempt to match the proposed commercial design type, but for which the number of stages and hence performance is more difficult to calibrate.

For at least one development, the mixer-settler approach was used to obtain design data, and a small pilot sieve tray extractor was deployed to demonstrate lack of fouling in continuous operation. Tray design for the 50 mm extractor was too crude to enable a detailed estimate of tray efficiency, but this was not a problem for design of the commercial unit [2, 3].

Summary

In summary, separations do not have to drive the scale of continuous pilot plants, unless the separation itself is the main issue to be de-risked in the development program. The separation devices chosen should be capable of generating reliable design data where needed, and may be operated in semi-continuous mode to examine separation factors, or diagnose potential for foaming, entrainment, or skin-temperature derived degradations. In this manner, modeling plus experiment in separations can work in tandem to reduce the require scope, cost, and timing for process development.

References

1. S. Sundaram and L. B. Evans, *Shortcut procedure for simulating batch distillation operations*, **Ind. Eng. Chem. Res., 1993, 32** (3), 511-518.
2. R. E. Treybal, **Mass Transfer Operations**, McGraw-Hill, NY (1980).
3. R. E. Treybal, **Liquid Extraction**, McGraw-Hill, NY (1963).

Figure 1: Relative cost of research, development, and demonstration phases

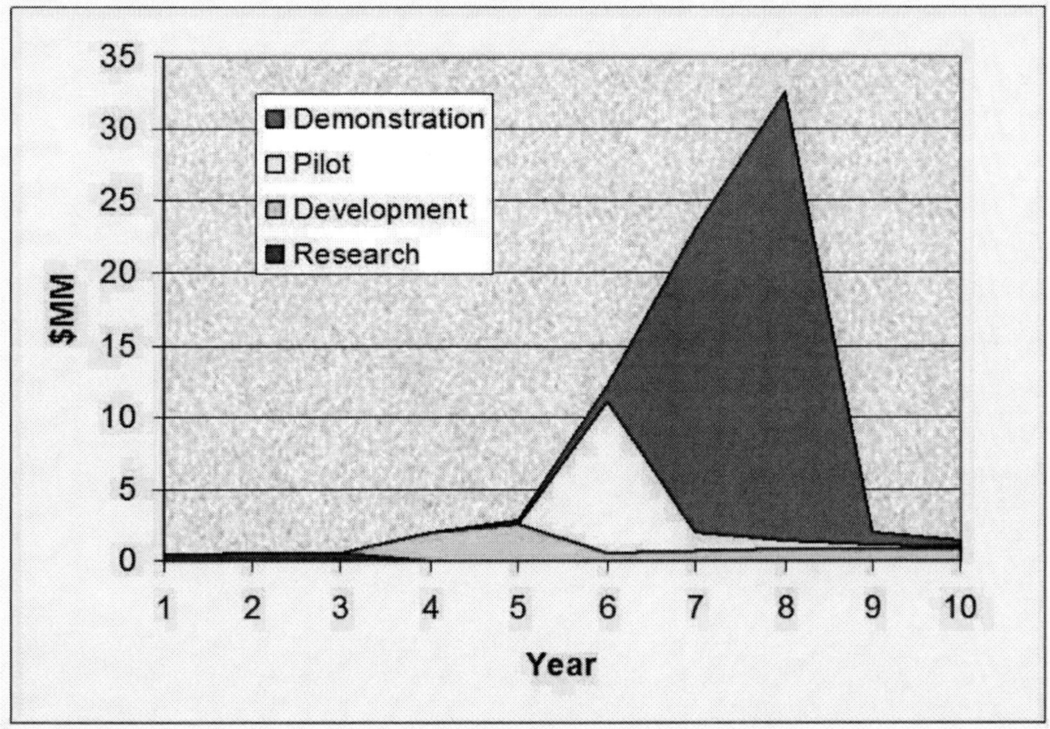

Figure 2: Batch distillation revealing reactive formation of volatile component from heavy ends

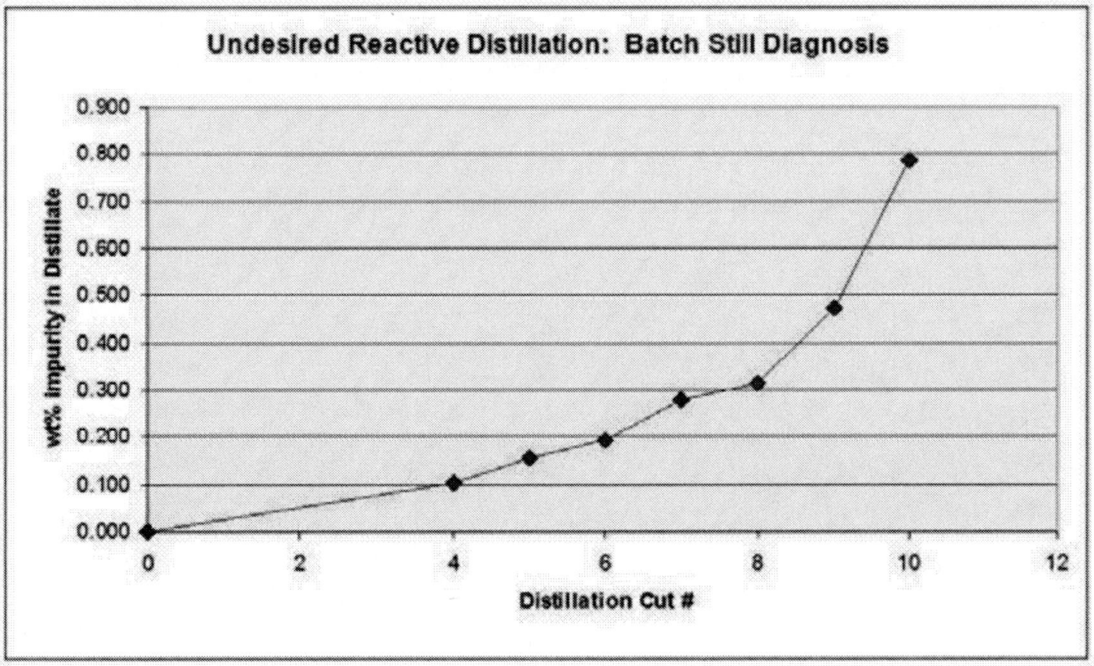

Figure 3: Inverted Distillation Sequence to Resolve Undesired Reactive Distillation

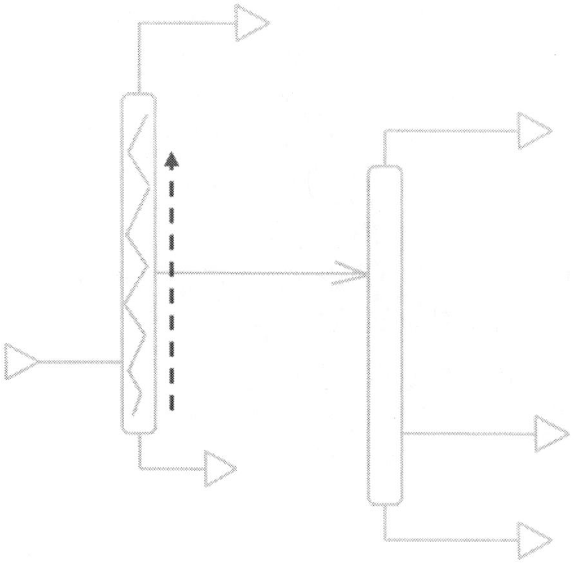

479574 Distillation: An Under-Plowed Field

Abstract:

We all know and agree that the heart of a chemical process resides in the reactor section of a chemical plant, where product forms. However, the majority of the invested capital goes to the unit operation separating desired product from by-product and solvent. The same unit operation also comprises the majority of the operating costs for a given chemical process.

Distillation is the most common separation unit operation in the chemical processing industry. It was studied intensively from the 1890's through the 1980's. However, since the early 1990's, distillation has become rather passe with regard to academic investigators. But, that response does not mean distillation is devoid of interesting problems to investigate.

Advances in computer power and user-friendly software opens new opportunities for investigating the fundamental features of distillation. In this presentation, we present a dynamic model for describing the separation that occurs on each tray of a laboratory-scale distillation unit. We then show how this model can be used to scale a distillation operation from laboratory-sized equipment to commercial-sized equipment.

PERFORMING OPTIMAL PILOT PLANT TESTS FOR DESIGN OF LIQUID-LIQUED EXTRACTION COLUMNS

Authors:

Donald J. Glatz
Brendan Cross
Thomas Lightfoot

Koch Modular Process Systems, LLC

Prepared for Presentation at

AIChE Spring Meeting
San Antonio, TX
March 2017

INTRODUCTION

The best and most reliable method for the design of a liquid-liquid extraction (LLE) column is by pilot plant testing with actual feed and then relying on known scale up correlations for the specific type of extraction column[1]. In general, the development procedure consists of the following steps; (1) selecting the solvent and generating liquid-liquid equilibrium data, (2) using this information and the project requirements to select the column type to use, (3) performing a pilot plant test to generate the data for accurate scale up to the production column, and (4) designing and subsequent fabrication of the column. This development procedure has been well documented in a previous paper[2].

The purpose of this paper is to focus on the pilot plant test method. During process development work, the ability to pilot plant test is often limited by a number of factors. Some of the key factors are:

- Availability of the feed materials. Often very little if any actual feed material is available.
- Known purity of the solvent. Since the solvent is a recycle stream, the solvent will generally not be pure during real commercial operation.
- Time restraints. Often testing must be done quickly to meet deadlines.
- Disposal of waste materials. Generally it is desirable to minimize the amount of waste material generated during pilot plant testing.

For all of these reasons it is important to have an optimized test method that can be completed in a short time frame using a minimal amount of feed and solvent. For basic R&D work, experimental design is a well proven and accurate method for process optimization. However, for pilot plant testing, this method requires a significant amount of feed and solvent and can take a long period of time. As presented in this paper, a Step-by-Step optimization method is outlined which saves both time and material. This method does require some experience but when it is applied correctly it can insure that an extraction column is designed correctly and performs as required.

LABORATORY WORK PRIOR TO PILOT TESTING

As mentioned in the Introduction, the first step for an extraction process is selecting the solvent and generating equilibrium data (distribution coefficient vs solute concentration). There are a number of methods for doing this. One such method was outlined in a previous paper[3]. In addition to the raw equilibrium data, observation of the phase behavior is just as important, especially for selecting the right type of column for scale up and developing a test plan for that column. The following observations are recommended:

- Mixing intensity required to generate good phase dispersion.
- Mass transfer rate/the time required to reach equilibrium. Generally we have learned that equilibrium can be reached in 2 minutes of high speed mixing, except if one or both phases are viscous (> 100 cps).
- After mixing, the time for the phases to separate and achieve a clear interface.

- The interface behavior – Is there a rag layer (solids) or an emulsion band that forms at the interface?
- Entrainment – Is some entrainment observed in one or both phases?

A typical data sheet for generating liquid-liquid equilibrium data is shown in Figure 1. A review of this sheet shows that for each mix-decant it will provide the material balance, the solute concentration in both phases, the cross solubility of the phases, and the observation of the phase behavior. With some experience, this data will help the engineer select both the best extraction column for the process and the starting point for pilot plant testing which will be discussed below.

PILOT TEST METHOD – STEP-BY-STEP OPTIMIZATION

A summary of the proposed method for pilot plant testing an agitated extraction column is provided here. This method requires fairly quick analysis of the samples generated in the pilot column, with the solute concentration in the raffinate usually being the key marker for the performance. It is also highly desirable to perform the test in a glass-shell extraction column so that the operators can observe the phase behavior and "tweak" the column operating conditions accordingly. The procedure is outlined as follows:

1. Select the continuous and dispersed phase for the process. There are a number of factors which can affect which phase is continuous and which phase is dispersed, but this is not the topic for this paper.

2. Select the column for scale-up and then set up the pilot column with multiple inlets. If possible it is best to set up a column that is expected to be taller than required. A manifold system can be included to vary the inlet location for the dispersed phase. The continuous phase must be fixed at the end where it enters, whereas the manifold system can be used to vary the inlet location for the dispersed phase.

3. Select the initial operating conditions for the test, i.e. the height of the column, the capacity, the solvent to feed ratio, the temperature and the initial agitation speed. Suffice to say that this is difficult to do without some previous experience with similar extraction processes. However, with some experience an engineer or scientist can usually determine acceptable starting conditions for a given type of extraction column.

4. Once the initial operating conditions are set, the column should be started up in the usual manner[4], and the agitation set at a low speed. This speed should be significantly below the flooding point for the process. Observation of the phase behavior in the glass shell column will determine this speed. The turnover time for the process should be calculated. Turnover time is the volume of the column divided by the feed + solvent rates. During the first two column turnovers, the agitation speed can be tweaked up slightly if warranted by observation of the phase behavior. However, after two turnovers, the agitation speed should remain fixed for the duration of the run which is a total of five (5) turnovers from the initial start-up. After five turnovers, raffinate and extraction samples are taken and analyzed for solute

concentration.

5. Next the agitation speed is increased slightly and then fixed for an additional three (3) turnovers before sampling and analyzing. This procedure is repeated, progressively increasing the agitation speed and sampling after three turnovers each time until flooding is observed in the column.

   NOTE: For most extraction process, we have learned that five column turnovers upon start up followed by three turnovers after making slight system modifications are sufficient to reach steady state. However, it is possible that for some systems, more turnovers may be required.

6. An evaluation of the results from this first series of tests will determine the next direction for the pilot plant testing. Usually the next direction is optimization of the solvent to feed ratio (S/F) vs the number of stages (height of the column). If the results from the first series of testing are good (the required solute extraction is achieved), then the next test runs should be at progressively lower solvent to feed ratios and/or lower height of the column. However, if the initial tests do not achieve the desired solute extraction, then the next test runs should be at progressively higher solvent to feed ratios and/or more agitated stages. Which way to go is generally determined by the expected economics for the process. Usually the more economical solution is to use more stages and thereby minimize the solvent usage.

7. After optimizing the S/F vs column height, the next variable to test is capacity, pushing the capacity to progressively higher and higher capacities to determine the optimal capacity for the process. Sometimes it is also desirable to test at lower capacity, especially if the production column is scheduled to begin operating at a capacity that will be significantly lower than the design capacity.

8. Finally, if necessary, additional variables such as temperature, pH and variable mixing intensity throughout the column (due to variable hydraulic behavior between the two phases from one end of the column to the other) can be tested.

If applied correctly, this test procedure can usually be completed in three operating days. Day 1 is finding the optimal agitation speed, Day 2 is optimizing S/F ratio vs column height, and Day 3 is optimizing the capacity A fourth day may be required if any other variables mentioned in Step 8 are tested, and/or multiple feeds are required. Two actual examples are now presented which demonstrate the Step-by-Step optimization procedure.

Example 1 – Carboxylic Acid Extraction

The application required the extraction of acetic acid from a whole fermentation broth using ethyl acetate as the solvent. The goal was to achieve >95% extraction of the acetic acid at the minimum solvent to feed ratio. As expected for fermentation broths, this extraction system was extremely prone to emulsification. As such the efficiency of an extraction column was expected to be low (high HETP) and only gentle agitation could be applied. The distribution coefficient for this process is approximately 0.9. Using the Kremser equation[5], the number of theoretical stages vs S/F are calculated as follows:

S/F = 2.0, number of theoretical stages = 3.9
S/F = 1.7, number of theoretical stages = 4.9
S/F = 1.5, number of theoretical stages = 6.1

Based upon these calculations plus observation of the slow phase separation and high tendency towards emulsification a 1" diameter x 12' plate stack height, glass-shell KARR® Column[6] was selected for the testing. The aqueous phase was selected as the continuous phase. The equipment set-up is shown in Figure 2. The results from three days of pilot plant testing are shown in Table 1. The progression of the pilot plant test was as follows:

- Runs 1 – 6 were performed on Day 1 at a broth feed rate of 83 cc/min, S/F = 1.7 weight basis, and room temperature. The starting point for Run 1 was 30 strokes per minute (SPM). This was increased to 40 SPM for Run 2 and 50 SPM for Run 3 where the column flooded. The desired acid extraction was achieved. Runs 4 and 5 with 8' plate stack height failed to achieve the desire acid extraction whereas 10' plate stack height in Run 6 did achieve the desired extraction. It was decided to conservatively use 12' plates stack height for all additional runs.

- Runs 7-9 were performed on Day 2 at the lower S/F ratio of 1.5. Initially the feed rate and solvent rates were adjusted to maintain the same combined flow rate as the previous runs, i.e. the new feed rate was 90 cc/min. The desired acid recovery was not achieved in Run 7. In Runs 8 and 9 the feed rate was adjust down to 83 and 65 cc/min respectively, while maintaining the S/F ratio of 1.5 and the desired recovery was achieved for both runs. It was decided to lock in the capacity at 83 cc/min and S/F of 1.5.

- During Day 3, the temperature was increased to 45 °C in the column. This proved to be an excellent operating temperature for this process. The operators reported that the phase behavior was more stable at this temperature. As shown in Run 11, successful operation at 50 SPM resulted in almost 99% acid extraction. This was determined to be the scale up run for the process.

Based upon these results, a 22" diameter KARR® Column was designed, built and has operated successfully for this process. The total amount of fermentation broth feed used for the testing was approximately 70 gallons.

Example 2 – Methanol Extraction

The application required the extraction of methanol from an organic stream using water as the solvent. The target was to reduce the methanol concentration in the organic phase from 2.5% to less than 0.1% using a specified amount of water (S/F = 0.32, weight basis). The liquid-liquid equilibrium data showed that 5-6 theoretical stages were required. A pilot KARR® Column was set up as shown in Figure 3. Due to the low S/F ratio (low water rate relative to the organic feed rate) the organic phase (light phase) was selected as the continuous phase and thus the interface was controlled at the bottom. Results from two days of pilot plant testing are shown in Table 2.

The test conditions and results are summarized as follows:

- All tests runs were made using a plate stack height of 6', i.e. the upper water inlet was never used. Also all runs were room temperature (20-25 °C). The first three runs were made using standard 2" plate spacing throughout the agitated zone, a feed rate of 150 cc/min and progressively higher agitation speed (SPM). The result show that the column flooded at 110 SPM, but had not achieved the desired alcohol concentration in the raffinate stream.

- Observation of the column during the first three runs showed that the dispersed phase water droplets would drop down through the column fairly rapidly, with little hold-up until they reached a point approximately 2' above the organic phase inlet. This was the point where flooding initiated during Run #3. Thus, the column was shut down and drained. The plate stack was removed and the spacing between the perforated plates was modified as follows; from the bottom of the plates stack 1' of 6" spacing, followed by 1' of 4' spacing and then the standard 2" spacing for the remainder of the plate stack. Thus, for any given agitation speed, significantly less power would be imparted into the bottom 2' of the agitated zone.

- The difference in the physical operation of the column was dramatic. As shown by the data, much higher agitation speeds were possible (without flooding) and the dispersed phase hold-up was observed to be significantly higher throughout the column. After Runs 4 and 5, the operators determined that higher capacity could easily be achieved with the modified plate spacing. As such for Runs 6-9, the feed rate was increased to 180 cc/min. The desired raffinate concentration was achieved at 200 SPM (Run 9). One additional run was made at a feed rate of 210 cc/min and the desired raffinate concentration was again achieved at 175 SPM.

Based upon these results, a 45" diameter KARR® Column was designed, built and has operated successfully for this process. The total amount of organic feed used for the testing was approximately 55 gallons.

ADDITIONAL OPTIMIZATION FROM PILOT TESTING

During pilot plant testing of liquid-liquid extraction processes, there are other process conditions that can be observed and optimized. In fact, pilot plant testing may be the only way some of these can be recognized and the correct design modification implemented. Some of the most common are listed here:

- Using the proper material of construction for the internals due to the wetting characteristics of the system. It is important for the continuous phase to preferentially "wet" the surface of the internals so that the dispersed phase does not coalesce on the internals. This is easily seen when using a glass shell extraction column.
- Minimization of entrainment – The amount of entrainment out each end of the column can usually be measured by immediate centrifuge of samples leaving the column. If the process

will not tolerate a certain level of entrainment, then modifications can include (1) expanding the disengaging chamber (2) installing a coalescing element in the disengaging chamber, or (3) building an external separator.

- Interface behavior – Does a solids "rag" layer or an emulsion band form at the interface. If so, how is this controlled?
- Co-extraction of impurities besides the desired solute – If this occurs, how can it be handled? This may require a "backwash" section in the column or using a fractional extraction technique[7].
- When flooding occurs, how is this handled in order to return the column to smooth operation? It is best to learn this during pilot plant testing so that it can be applied commercially in the future.

CONCLUSION

For most liquid-liquid extraction processes it is possible to develop and perform a pilot test plan that will save time and money but still generate the data for accurate design of a production column. The generalize procedure was outlined in this paper and supported by two real examples where this method was applied.

REFERENCES

1. Cusack, R., Karr, A., "A Fresh Look at Liquid-Liquid Extraction, Part 3, Extractor Design and Specification", Chemical Engineering, April 1991

2. Cusack, R., Glatz, D., Apply Liquid-Liquid Extraction to Today's Problems", Chemical Engineering, July 1996

3. Glatz, D., Parker, W., "Extraction Enhancement, One Step at a Time", Chemical Engineering, November 2004

4. Glatz, D., Lightfoot, T., "Using Extraction Columns for Recovery of Valuable Products from Biomass", AIChE National Meeting, November 2012

5. Cusack, R., Fremeaux, P., and Glatz, D., "A Fresh Look at Liquid-Liquid Extraction, Part 1, Extraction Systems", Chemical Engineering, February 1991

6. Cusack, R., Fremeaux, P., "A Fresh Look at Liquid-Liquid Extraction, Part 2, Inside the Extractor", Chemical Engineering, March 1991

7. Glatz, D., Lightfoot, T., "Designing Columns for Fractional Liquid-Liquid Extraction", AIChE National Meeting, November 2010

Figure 1 - Shake Test Data Sheet

Client: **Application:**

Description: **Feed:**

Solvent:

	Description		Weight (grams)				Separation		Solute Analysis			
No.	Feed	Solvent	Feed	Solvent	Raff.	Ext.	Time	Observation	Feed	Solvent	Raff.	Ext.

Figure 2 – Pilot Plant KARR Column Set-Up
Acetic Acid Extraction from Broth using Ethyl Acetate

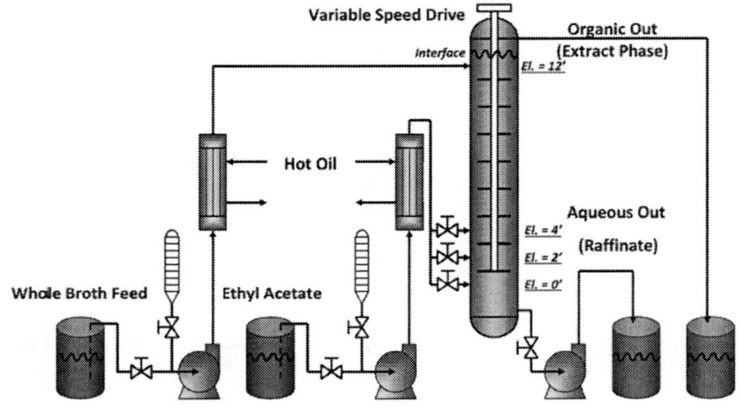

Figure 3 – Pilot Plant KARR Column Set-Up
Extraction of Alcohol from Acrylate Using Water

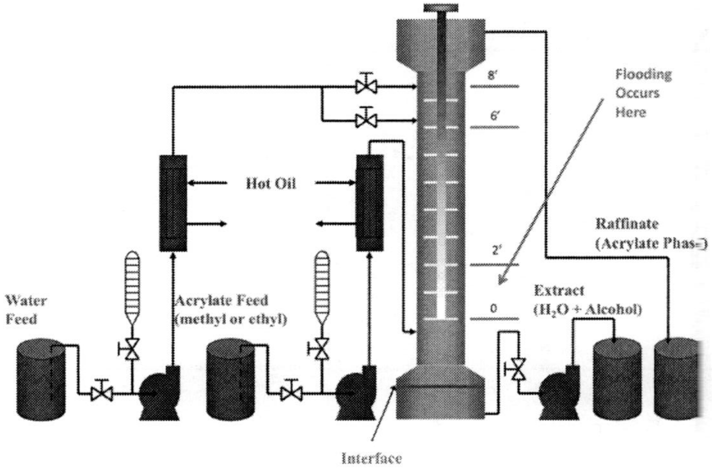

Table 1 - Pilot Plant Test Results
Acetic Acid Extraction from Broth using Ethyl Acetate

Run #	Agitated Height (ft)	Feed Rate (cc/min)	Temp. C	S/F Ratio	Agitation SPM	% Acid Recovery
1	12	83	23	1.7	30	95.0
2	12	83	23	1.7	40	97.5
3	12	83	23	1.7	50	Flooded
4	8	83	23	1.7	30	93.3
5	8	83	23	1.7	40	Flooded
6	10	83	23	1.7	30	96.3
7	12	90	23	1.5	30	94.3
8	12	83	23	1.5	30	96.5
9	12	65	23	1.5	40	97.6
10	12	83	45	1.5	30	98.3
11	12	83	45	1.5	50	98.7
12	12	83	45	1.5	60	Flooded

Data generated in 25mm diameter KARR Column
Total amount of fermentation broth used = approximately 70 gallons

Table 2 - Pilot Plant Test Results
Extraction of Methanol from Acrylic Feed Stream Using Water as the Solvent

Run #	Plate Stack	Feed Rate (cc/min)	Water Feed Rate (cc/min)	Agitation Speed (SPM)	Raff. Conc. % Alcohol Target <0.1%
1	1	150	45	75	0.165
2	1	150	45	100	0.124
3	1	150	45	110	**FLOODED**
4	2	150	45	110	0.169
5	2	150	45	140	0.112
6	2	180	54	100	0.203
7	2	180	54	125	0.146
8	2	180	54	150	0.118
9	2	180	54	200	**0.078**
10	2	210	63	175	**0.084**

Notes: KARR Column with 1" diameter x 6' plate stack height
Plate Stack #1 has constant 2" plate spacing throughout
Plate Stack #2 from bottom: 1' of 6", 1' of 4", and 4' of 2" spacing between plates
Total Amount of organic feed used = 55 gallons

483476 Smart Manufacturing Business and Technology Practice

Abstract:

Smart Manufacturing (SM) uses the integration of next generation Operations Technology (OT) and Information Technology (IT) to realize significant untapped market opportunities. Opportunities accrue from a broader, more customized product space and accelerated dynamic, and precision manufacturing, drawing upon the cumulative impacts of every step across entire enterprise value and supply chains. Significantly increased energy, material and workforce productivity and improved environment sustainability are pathways to opportunities as well as outcomes of enterprise optimization. Smart Manufacturing (SM) is a term of 'practice' to just technology. It is in what way, at what speed, and how well data and next generation IT are applied that impact the future of individual manufacturers and their supply chains. Smart Manufacturing (SM) is therefore the opportunity-based and business oriented application of next generation advanced sensors, controls, platforms and modeling. Building meaning through a lens of practices that orchestrate capabilities to achieve business and operational objectives makes it possible to describe SM itself as set of practices that distinguish it from today's 40 year history of applying IT to manufacturing.

482647 Process Intensification for Reactions and Reactors

Abstract:

We will describe various concepts for intensifying catalytic reactions and reactors, including bifunctional materials and cascade reactions, fairly mature by now microtechnology, hybrids of reactors and separations, and multifunctional reactors including static micromixers and microheat exchangers (spatial domain), periodic operation (temporal domain), and use of alternate energy forms such as microwaves. These concepts will be elucidated with examples taken from biomass, shale gas, and ammonia chemistry.

483477 Process Intensification at Different Length-Scales

Abstract:

Process intensification (PI) concepts offer opportunities to enhance the energy efficiencies, and reduce the emissions and costs for the manufacture of chemicals, medicines and fuels. This talk will describe our use of PI concepts at different length-scales to design and demonstrate two process systems: a highly efficient, thermally integrated system for the production of hydrogen, and a nanoscale cascade catalytic system for the hydrogenation of CO_2 to value-added products. The hydrogen production system is designed to use latent from the vaporization of H_2O to control the axial temperature gradient of a reactor that couples autothermal reforming, water gas shift and preferential oxidation. Cascade systems use multiple catalytic species to facilitate multi-step reactions in a single vessel. A low temperature, heterogeneous cascade system is demonstrated for the selective hydrogenation of CO_2 to methanol and dimethyl ether via formic acid and methyl formate intermediates. Technoeconomic analyses indicate that these two modular systems could reduce the cost of distributed hydrogen production and CO_2 hydrogenation.

479998 Using Novel Multiphase Simulation Techniques for Innovative Process Development - Case Studies in PO Processes

Abstract:

Multiphase simulations have been the topic of extensive research and developments giving deep insights and improved understanding of phenomenon observed in a wide variety of chemical process industry. Their application to polymerization process has been also widely reported but the long road to full understanding in industrial terms remains full of potholes.

Borstar™ PE or PP processes include a combination of (i) slurry loop and (ii) gas-phase fluidized-bed reactor(s) for polymerizing ethylene or propylene. Both kinds of reactors operate in a multiphase regime with either liquid or gas in presence of growing particles that range from microns to millimeter sized particles. For taking advantage of the Borstar™ process, it becomes vital to generate meaningful cause-and-effect understanding of particle-related phenomenon that occur in each kind of the reactor and on the how the polymer properties are developing with time & space.

Standard DPM & TFM approaches for modelling either of these polymerization reactors may be sufficient for understanding fluid-particle interactions at limited scales, obtaining simulation results on industrial sized reactors requires substantial computation demands. In addition, including polymerization mechanisms in order to perform 'warm' reactor simulations remains an even larger challenge.

Borealis has been working with academic partners who have made significant progress in how state-of-the-art multiphase models can be extended to allow understanding of fluid-particle segregation for large simulation geometries in reasonable time frame and incorporate reaction mechanisms. By carrying out such 'warm' reactor simulations Borealis has been able to obtain detailed information on the polymer particle characteristics. This information has helped us further optimize the operation of our reactors and enhance their operating window. Consequently, the availability of our polymerization reactors is greatly improved and led to optimization in scheduling.

Two separate case studies dealing with large scale simulations of multiple-leg loop reactors as well as gas-phase fluidized bed reactors will be shared to demonstrate the advantage such approach provides.

480180 Composition Broadening in Polymerization Systems

Abstract:

Because of their stochastic nature, polymerization processes generate mixtures of chains that differ from each other in their molecular architecture. One should not expect chains with the same number of repeating units, for example, even when process conditions remain the same across both time and space. These differences are mostly determined by the interaction among the various steps of the polymerization mechanism. Other factors play important roles too, with two of the most important additional factors being reactor configuration and mixing. Our work considers the impact all of these effects on polymer composition.

Polymer composition is a key quality characteristic. The production of Linear Low Density Polyethylene (LLDPE) illustrates its importance. In this system, a-olefins like butene, hexene, and octene introduce small side chains into the ethylene based polymer backbone. These side chains reduce the density and the crystallinity of the final polymer, a desirable characteristic for certain applications. Increasing levels of a-olefins can also lead to products with lower molecular weights, since a-olefins are typically not as reactive as ethylene towards chain growth. The strong dependency of key product properties on the level of incorporation of a-olefins comes at a cost, however, since missing a target by adding too much or too little of these comonomers can lead to products with unexpected properties.

Tadmor and Biesenberger[1] already looked at differences in molecular weight distributions using continuous tank reactors with both perfect micromixing and complete segregation. Their intention was to illustrate how incomplete mixing affects the spread of those distributions in reactors that would be homogeneous otherwise. Mecklenburgh[2] used a Free Radical polymerization system to study the impact of mixing in various reactor configurations. The study concluded by stating that copolymerization distribution curves should provide insight into the mixing patterns inside a reactor and lamented on the lack of analytical techniques to validate this hypothesis. In a more recent attempt at measuring micromixing, Atiqullah and Nauman[3] applied gradient elution thin-layer chromatography to the styrene - methyl methacrylate system. They developed an extension of previous lamellar stretch models that included backmixing effects, but concluded that their reactor was too small for them to observe the impact of micromixing on copolymer composition distributions. The impact of imperfect mixing on polymer properties has been addresses by various authors by relying on compartmental models, a good example of this being the work of Marini and Georgakis[4]. In this work, polydispersity index was used as a measure of quality of Low Density Polyethylene (LDPE), and the effect on this property was studied for different reactor configurations and process conditions. This work was later reviewed and analyzed in more detail, with particular emphasis on compartmental models, by Zhang and Ray[5].

Using examples in Free Radical polymerization, we rely on computational tools to study the spread in the polymer composition distribution under the influence of the factors mentioned earlier. Our work shows how critical comonomer reactivity is and how important it is to control all factors in order to generate a product with the desired composition distribution. Comonomer reactivity is critical because wide residence time distributions lead to differences in conversion and, as a result of this, differences in composition of the unreacted monomer mixture. These differences in turn lead to wider polymer composition distribution. In extreme cases, multimodal distributions are even possible. Our work uses simple geometries to illustrate these problems and demonstrates that product quality is highly sensitive to what most would consider simple details in polymerization reactor design.

---

[1] Tadmor, Z.; Biesenberger, J. A. Influence of Segregation on Molecular Weight Distribution in Continuous Linear Polymerizations. *I&EC Fund.* **1966**, 5, 336-343

[2] Mecklenburgh, J. C. The influence of mixing on the distribution of copolymerization compositions. *Can. J. Chem. Eng.* **1970**, 48, 279-285

[3] Atiqullah, M.; Nauman, E. B. A Model and Measurement Technique for Micromixing in Copolymerization Reactors. *Chem. Eng. Sci.* **1990**, 45, 1267-1279

[4] Marini, L.; Georgakis, C. The Effect of Imperfect Mixing on Polymer Quality in Low Density Polyethylene Vessel Reactors. Chem Eng Commun. 1984, 30, 361-375

[5] Zhang, S. X; Ray, W. H. Modeling of Imperfect Mixing and Its Effects on Polymer Properties. *AIChE J.* **1997**, 43, 1265-1277

481165 Tracking Down Organic Acids and Sulfides from Petroleum Fluids: An Application of Real Molecule-Based Crude Assay Characterization and Segment-Based PC-SAFT

Abstract:

As global crudes being depleted, more interest and concern have been shown on heavy crudes with larger total acid number (TAN) and higher sulfur content. To process such crudes, aggressive corrosive attack should be expected at elevated temperature (300 - 350 °C)[1]. Naphthenic acid (NAPA) corrosion is known to be a non-aqueous corrosion process, which may occur inside units including pipelines and distillation columns. Sulfides on the other hand, are also corrosive, but may offset NAPA's corrosion to some extent by forming metal sulfide scales on the equipment surfaces. However, neither TAN nor total sulfur content correlates well with corrosivity of engineering alloys.[2] The present work provides specific molecular structure and composition of NAPA and sulfides by applying real molecule-based characterization to the crude assay. Furthermore, segment-based PC-SAFT as an equation of state is used to calculate vapor-liquid equilibrium of the petroleum, and simulate atmospheric distillation column to precisely track down the molecules of interest.

1. Bota, G. M.; Qu, D.; Nesic, S.; Wolf, H. A., Naphthenic Acid Corrosion of Mild Steel in the Presence of Sulfide Scales formed in Crude Oil Fractions at High Temperature. In NACE International: 2010.
2. Dettman, H. D.; Li, N.; Luo, J., Refinery Corrosion, Organic Acid Structure, and Athabasca Bitumen. In NACE International: 2009.

483079 Process Intensification in Multicomponent Distillation

Abstract:

Process intensification (PI) is an emerging concept which describes the design innovations that lead to significant shrink in size and dramatic boost in efficiency in a process plant. Distillation, which accounts for 90-95% of all separations in a chemical plant and ~3% of the world's energy consumption, is therefore a crucial component in PI. In the context of multicomponent distillation, we believe that PI stands for minimizing the number of pieces of equipment as well as total cost while maintaining high efficiency through innovative process synthesis. This will lead to compact, easy-to-operate, energy efficient and cost effective processes. Here, we consider several different levels of PI in multicomponent distillation, including: 1) Introducing thermal couplings that eliminates some or all reboilers and condensers to save energy and reduce capital cost; 2) Using thermal couplings strategically by identifying places where heat exchangers that use intermediate temperature utilities should be kept to enhance thermodynamic efficiency, as well as places where thermal couplings are preferred in order to provide substantial energy savings but incur no or minimal penalty in utility; 3) Performing simultaneous heat and mass integration among thermally coupled distillation columns to reduce the number of columns and heat duty requirement; 4) Conducting any thermally coupled distillation in only one single column shell using the form of a dividing wall column that is fully operable; 5) Being able to systematically derive numerous configurations that use 1 to (−2) column shells from a traditional (−1)-column configuration. Here, we demonstrate these aspects of PI through examples, illustrating the importance of PI in multicomponent distillation and how this concept can be utilized by process engineers to design cost effective and energy efficient distillation systems.

482646 Autothermal Process Intensification: Overcoming Heat Transfer Bottlenecks in Non-Equilibrium Chemical Processes

Abstract:

Heat transfer is the bottleneck in many chemical processes, limiting throughput for a reactor even when chemical reaction and mass transfer are very fast. Heat transfer can be enhanced through a number of strategies, often adding significantly to the capital cost of a project. However, as long as heat transfer is rate limiting, reactor capacity only scales as the square of reactor diameter compared to the cube of reactor diameter for volumetric chemical rate-limited reactors. This not only constrains throughput as reactors are built larger, but substantially eliminates the economic advantage of building them larger.

The heat transfer bottleneck can be overcome through *autothermal chemical processing*, which balances the energy demand of endothermic chemical reactions with energy supply from exothermic chemical reactions within an adiabatic chemical reactor. Although autothermal chemical reactors have been previously demonstrated for *equilibrium* chemical processes (prominent examples include methane steam reformers and biomass/coal gasifiers), we are developing it for inherently *non-equilibrium* processes. Identifying complementary endothermic and exothermic reactions that can occur simultaneously while preserving the same products of an externally heated (or cooled) chemical process is not trivial. However, the gains in *process intensification* would be substantial, increasing throughput and reducing capital and operating costs of a chemical process.

We have applied the principle of non-equilibrium autothermal chemical processing to the pyrolysis of biomass. Although the enthalpy for pyrolysis is relatively small (comparable to the enthalpy of evaporation for methanol) and operates at 500°C, throughput for a pyrolyzer is constrained by the rate that thermal energy can be transported into the reactor. This energy is conventionally supplied from hot flue gas conveyed through heat exchanger tubes or hot granular media like sand or steel shot mixed directly with the pyrolyzing biomass. Although these heat transfer systems add tremendous complexity to the pyrolyzers, they can sustain only modest thermal fluxes in practical pyrolysis systems, limiting the commercial prospects for pyrolysis energy systems.

Autothermal pyrolysis is achieved by admitting a small amount of oxygen to a fluidized bed reactor, which partially oxidizes biomass reactant or pyrolysis products to provide the enthalpy of pyrolysis. Experiments performed in a 15.4 cm dia. fluidized bed reactor capable of operating as either a conventional or autothermal pyrolyzer found no significant loss in bio-oil yield or quality for these two modes of operation at comparable biomass throughput. More importantly, by removing the heat transfer bottleneck of conventional pyrolysis, operation in autothermal mode allowed biomass throughput to be dramatically increased, reaching five times the capacity of the conventionally operated pyrolyzer. Our analysis indicates oxidation of non-condensable gas species released during pyrolysis provided about half the energy for pyrolysis while the remainder of the energy is thought to come from partial oxidation of lignin in the biomass and char in the pyrolysis products.

482648 Process Intensification and Modular Production for Converting Stranded Gas to Value-Added Products

Abstract:

In the United States and rest of the world, there are abundant gas resources which are either physically or economically stranded. These gases include flaring gas in shale oil field, refinery off-gas, coal-bed methane, shale gas in disadvantaged locations. Upgrading of these gases to value-added liquid products could reduce the demand on crude oil by about 20% in the United States.

On-site modular production of value-added products from stranded gases become attractive. Due to economy of scale, modular production typically doesn't compete well with conventional large scale production. As a result, process intensified modularity becomes important because it will enhance productivity, energy efficiency and flexibility that allows the modular production to be economically viable. What is process intensified modular production? It consists of novel apparatuses and techniques that, compared to those conventionally practiced today, are expected to bring dramatic improvements in manufacturing and processing, substantially decreasing the ratio of equipment size to production capacity. Most importantly, the process intensification should reduce energy consumption, waste production, and ultimately resulting in cheaper and sustainable technologies. Illustrated in this presentation are synergistic integration of the following unit operations: reactor design, heat management and separation. Process intensification sets the foundation for small modular production to achieve both energy efficiency economic production.

Indirect conversion of natural gas to liquid products (GTL) via syngas has been commercialized but it requires huge capital investment, varying from $15 to 20 billion for a 100-140 bbl/d plant. Direct conversion of natural gas without going through syngas route has been under investigation in the past 5 decades, but no commercial processes are practiced to date. A large number of studies have been published on the subject over the past 50 years.

This study emphasizes the direct conversion of short chain parafins (C_1-C_3) into aromatics and olefins using transition metal promoted ZSM-5 zeolite catalysts. Catalyst activity, selectivity, deactivation and regeneration of metal-promoted ZSM-5 zeolite catalysts will be discussed. We will introduce a new approach that employs low temperature plasma to intensify catalytic reaction for natural gas conversion. Under low reaction severity, this approach synergistically integrates plasma reaction chemistry with novel heterogeneous catalysis that decouples methane activation from catalytic surface reaction, shifting rate-determining step from methane activation (cracking C-H bond) to surface C-C formation.

In summary, this presentation illustrates experimental research on direct non-oxidative conversion of natural gas via process intensified production. The challenge in advance the fundamental science aspects presented in direct natural gas conversion is discussed. By comparing with commercial hydrocarbon conversion processes, the industrial perspective on direct methane conversion is illustrated. Specifically, the strategies on the selection of reactor configuration, heat management and catalyst regeneration are highlighted. These commercialization strategies along with technoeconomic analysis are critical for direct natural gas conversion to value-added chemicals.

Linear Dynamic Model Identification and Data Reconciliation using Dynamic Iterative PCA (DIPCA)

Vipul Mann, Arun K. Tangirala, Shankar Narasimhan

Department of Chemical Engineering
Indian Institute of Technology Madras, Chennai, Tamil Nadu, India

Abstract

Identification of input-output models from data is of utmost relevance in chemical process industries and has applications in process monitoring, control and fault diagnosis. Input-output data used in such identification exercises often has measurement errors in both the variables. Model identification under such conditions translates to solving an errors-in-variables (EIV) problem which is difficult to solve using classical system identification techniques. A recently proposed method - Dynamic Iterative Principal Component Analysis (DIPCA) uses PCA framework to identify the process order, delay, model parameters, and error variances. DIPCA, however, has certain shortcomings under small sample conditions which limit its practical applications.

In this work, we address these shortcomings, namely ambiguity in order determination under small sample cases and arbitrary selection of stacking lag which leads to sub-optimal parameter estimates. We define a metric called 'd-selective eigenvalue ratio', or *d-SEVR* that sharply identifies the true order even for small sample cases. We also demonstrate the existence of an optimal stacking lag corresponding to the lowest error in estimation of error-covariance matrix. Finally, we use the identified model to obtain reconciled estimates of variables using Kalman Filter.

Keywords: Identification; Principal Component Analysis; Errors-in-Variables; Kalman Filter

1. Introduction

While performing data generation exercises in a plant, measurements are often corrupted with errors which could incur due to instrumentation error, observational error, situational error, or any other unknown source of error. The variances of such errors across both input and output variables are usually unequal i.e. the errors are heteroskedastic in nature. In the process industry, such data is predominantly used for optimizing control loop performances, making predictions about a given process variable, and monitoring process operations - all of which have model identification at their core. Therefore, it becomes extremely important to identify the underlying data generating process from erroneous measurements of the plant variables without any prior knowledge about the process.

Due to the presence of errors in both the variables, model identification under these conditions translates to solving an errors-in-variables (EIV) problem and it gets difficult to apply classical system identification techniques which are based on the assumption of error-free inputs. There exist different techniques for solving the EIV problem but the total least squares (TLS) formulation gives the most efficient solution. In [1], a way to use the instrumental variable techniques is proposed to solve EIV regression problem. On the other hand, multivariable statistical data analysis techniques have proven themselves to be highly successful in finding out linear relations between variables corrupted with measurement errors. Principal

Component Analysis (PCA) is one of the most widely used techniques for solving the TLS problem because of its simplicity, numerical robustness and computational efficiency. While PCA has widespread use in dimensional reduction, fault detection, and feature extraction, we focus on its application to solving the regression problem.

The generic formulation of PCA, which we call static PCA, is suitable for identifying steady-state or instantaneous linear relationships between variables. Dynamic PCA or DPCA was proposed in [2] as a natural extension of PCA by constructing the data matrix using stacking lagged versions of input-output variables and applying PCA treating them as separate variables. DPCA requires the order to be known exactly which is seldom known in reality. The general practice is to use heuristics based methods for order determinations, for example [3] and [4]. But, the major issue with DPCA is that it gives biased estimates of model parameters under heteroskedastic error conditions. Therefore, in order to get unbiased estimates of model parameters, the data has to be scaled appropriately using the error-covariance matrix as proposed in [5]. But again, the error-covariance matrix is rarely available in reality and has to be estimated from data. In [6], an iterative procedure called iterative PCA or IPCA is proposed for estimating the error-covariance matrix by solving a maximum likelihood problem in the static case. IPCA also estimates the number of constraints simultaneously. An extension of IPCA to the dynamic case is recently proposed in [7], which estimates the true model order, delay, and the underlying model parameters along with the input-output error-variances under large sample conditions.

In this work, we focus on DIPCA and attempt to study this algorithm in detail, address some of its shortcomings that arise due to small sample conditions, and explore its applications in data reconciliation.

The rest of the paper is organised in the following manner - In Section 2, we formally define our problem and the framework that we will be working under. In section 3, we review some of the algorithms necessary to gain a better understanding of this algorithm, namely static PCA, dynamic PCA, and iterative PCA. In Section 4, we explain the underlying algorithm in DIPCA and motivate it with an example. In Section 5, we highlight the shortcomings of this algorithm and propose techniques subsequently to address them in Section 6. In Section 7, we demonstrate the application of DIPCA in solving a data reconciliation problem, and present our concluding remarks in Section 8.

2. Problem formulation and objectives

Given a linear time-invariant (LTI), single input single output (SISO) system,

$$y^\star[k] + \sum_{i=1}^{n_a} a_i y^\star[k-i] = \sum_{j=D}^{n_b} b_j u^\star[k-j], \tag{1}$$

and input-output data corrupted with white-noise errors as

$$u = u^\star + e_u \quad e_u \sim \mathcal{N}(0, \sigma^2_{e_u}) \tag{2}$$
$$y = y^\star + e_y \quad e_y \sim \mathcal{N}(0, \sigma^2_{e_y}) \tag{3}$$

where u^*, y^* are noise free versions of the input and output variables, respectively, the objective is to identify the above dynamic difference equation model by estimating the process order $\eta \triangleq \max(n_a, n_b)$ (n_a is order of the output & n_b is order of the input), delay D, parameters $a_1, a_2, \dots a_{n_a}$, b_D
, $b_{D+1}, \dots b_{n_b}$, and the error variances $\sigma^2_{e_y}$ & $\sigma^2_{e_u}$.

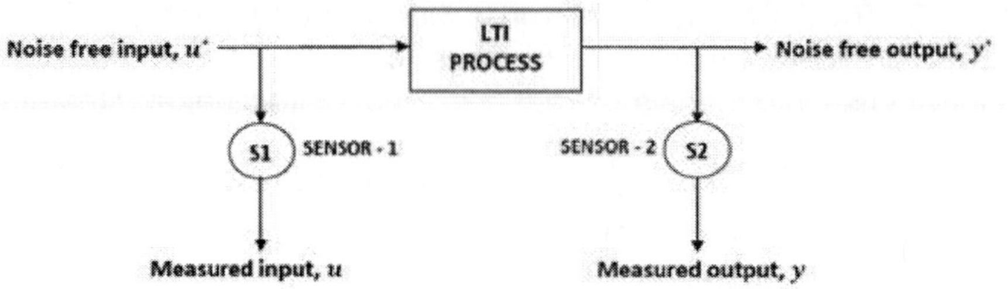

Figure 1: Process flow diagram depicting the data collection process

3. Foundations

The following subsections provide a brief overview of static PCA, dynamic PCA and iterative PCA from the viewpoint of model identification under the TLS framework.

3.1. Steady-state model identification using PCA

Suppose $\mathbf{x}[k] \in \mathbb{R}^{M\times 1}$, at any instant k, are related through d linear constraints as

$$\mathbf{A}\mathbf{x}[k] = \mathbf{0} \qquad A \in \mathbb{R}^{d\times M}, \text{ rank}(\mathbf{A}) = d \tag{4}$$

We can, therefore, say that the constraint matrix, $\mathbf{A}$, lies in the null-space of $\mathbf{X}$, and the model identification problem is equivalent to finding a basis for $\mathbf{A}$. $\mathbf{A}$ can be obtained through eigenvalue decomposition of the sample covariance matrix $\frac{1}{N}\mathbf{X}\mathbf{X}^{\mathbf{T}}$, of $\mathbf{X}$ where $\mathbf{X} \in \mathbb{R}^{M\times N}$ corresponds to N observations of $\mathbf{x}[k]$. For the noise free case, the last d eigenvalues will be identically 0 and the corresponding eigenvectors provide a basis for the null space of $\mathbf{X}$ i.e.

$$\sigma_{M-d+1} = \sigma_{M-d+2} = \ldots = \sigma_M = 0$$
$$\mathbf{A} = \begin{bmatrix}\mathbf{v}_{M-d+1} & \mathbf{v}_{M-d+2} & \ldots & \mathbf{v}_M\end{bmatrix}^T$$

For the noisy case, however, the last d eigenvalues does not exactly equal 0 and the extent of deviation from 0 will depend upon the error-variances in the variables. Under such conditions, heuristic based are used in order to determine the value of d. Under homoskedastic error conditions, we have

$$\sigma_{M-d+1} = \sigma_{M-d+2} = \ldots = \sigma_M = \sigma_{e_u}^2 = \sigma_{e_y}^2$$
$$\mathbf{A} = \begin{bmatrix}\mathbf{v}_{M-d+1} & \mathbf{v}_{M-d+2} & \ldots & \mathbf{v}_M\end{bmatrix}^T$$

But, when the error variances are heteroskedastic in nature with $\sigma_{e_u}^2 \neq \sigma_{e_y}^2$, PCA gives sub-optimal results. Also, satic PCA is limited only to identifying static relationships between variables and cannot discover dynamic relationships.

3.2. Dynamic model identification using Dynamic PCA

Dynamic PCA or DPCA is an extension of static PCA to the dynamic case which discovers dynamic relationships between a given set of input-output variables. The idea is to stack lagged versions of the variables up to a lag L in order to construct the data matrix $\mathbf{Z}$ as:

$$\mathbf{Z} = \begin{bmatrix} y[k] & y[k-1] & \ldots y[k-L] & u[k] & u[k-1]\ldots & u[k-L] \\ y[k+1] & y[k] & \ldots y[k-L+1] & u[k+1] & u[k]\ldots & u[k-L+1] \\ \vdots & & & & & \vdots \\ y[N-L] & y[N-L+1] & \ldots y[N] & u[N-L] & u[N-L+1]\ldots & u[N] \end{bmatrix}$$

Dynamic relationships between the independent and dependent variables can be obtained by modelling static constraints between these lagged variables.

Now, suppose $z = \begin{bmatrix} z_D & z_I \end{bmatrix}^T$ where, $z_D \in \mathbb{R}^{n_y \times 1}$ corresponds to the dependent variable y, and $z_I \in \mathbb{R}^{n_u \times 1}$ corresponds to the independent variable, u. Correspondingly, assume that $\mathbf{A} = \begin{bmatrix} \mathbf{A}_D & \mathbf{A}_I \end{bmatrix}$. Then, applying PCA on $\mathbf{Z}$ and considering the eigenvector corresponding to the smallest eigenvalue, the model is given by

$$\mathbf{A}\mathbf{Z}^T = 0 \implies \mathbf{z}_D = \mathbf{B}z_I \qquad \text{where, } \mathbf{B} = -\mathbf{A}_D^{-1}\mathbf{A}_I$$

However, in order to apply DPCA, the order η of the process has to be known a priori as DPCA involves exact stacking with $L = \eta$. Also, DPCA gives optimal parameter estimates only under homoskedastic-error conditions.

3.3. Iterative PCA for error-variance estimation

The underlying idea in IPCA is to perform PCA on a scaled version of original data with a scaling factor of $\Sigma_e^{-1/2}$, i.e.

$$z[k] :\rightarrow \boldsymbol{\Sigma}_e^{-1/2} z[k] \tag{5}$$

where, $\boldsymbol{\Sigma}_\mathbf{e}$ is the error-covariance matrix. Scaling the data matrix $\mathbf{Z}$ with $\Sigma_e^{-1/2}$ resolves this issue because it transforms the heteroskedastic-error problem into a homoskedastic-error problem. The homoskedastic-error problem can then be solved easily using dynamic PCA if the system order (η) is known. However, as a result of scaling, eigenvalues of the sample covariance matrix $\mathbf{S_z}$ are shifted by unity, i.e.

$$\lambda(\mathbf{S}_{\mathbf{z}_{scaled}}) = \lambda(\mathbf{S_z}) + 1$$

Hence, instead of zero eigenvalues, the eigenvectors corresponding to unity eigenvalues will give us the linear relations $\hat{\mathbf{A}}$ between variables. Also, the number of such unity eigenvalues gives us the number of constraints, d.

Iterative PCA iteratively estimates the noise covariance matrix Σ_e by solving the following maximum likelihood estimation problem,

$$\min_{\Sigma_\mathbf{e}} \quad N \log |\hat{\mathbf{A}}^{(\mathbf{i})}\boldsymbol{\Sigma}_\mathbf{e}(\hat{\mathbf{A}}^{(\mathbf{i})})^\mathbf{T}| + \sum_{i=1}^{N} (\mathbf{r}_i^T[k](\hat{\mathbf{A}}^{(\mathbf{i})}\boldsymbol{\Sigma}_\mathbf{e}(\hat{\mathbf{A}}^{(\mathbf{i})})^\mathbf{T})^{-1}\mathbf{r}_i[k]) \tag{6}$$

which is based on the assumption that the residuals generated as

$$\mathbf{r} = \hat{\mathbf{A}}\mathbf{Z} \tag{7}$$

are Gaussian-distributed with zero mean and variance $\boldsymbol{\Sigma}_\mathbf{e}$. Therefore, IPCA estimates $\mathbf{A}$ and $\boldsymbol{\Sigma}_\mathbf{e}$ alternatively in an iterative manner given an identifiability constraint is satisfied.

4. Dynamic Iterative PCA

Dynamic Iterative PCA or DIPCA, proposed in [7] is an extension of IPCA for the dynamic case where dynamic relationships between the input and output variables have to be identified. DIPCA is to IPCA what DPCA is to PCA.

4.1. *DIPCA algorithm*

The idea behind DIPCA is to use IPCA for estimating the error-covariance matrix Σ_e with a modification that Σ_e is a diagonal matrix comprising of σ_u^2 and σ_y^2. The order η is derived from the number of linear relations d as $\eta = L - d + 1$ since stacking in excess results in increased number of constraints. Once we know these quantities, we can then estimate the difference equation model $\mathbf{A}$ using PCA.

4.2. *Motivating example for DIPCA*

Consider a data generating process,

$$\text{P1:} \quad y^{\star}[k] + 0.4y^{\star}[k-1] + 0.6y^{\star}[k-2] = 1.2u^{\star}[k-1]$$

Order η, delay η_u, error-covariance matrix Σ_e are all assumed to be unknown and L is arbitrarily chosen to be equal to 5. The error-variances in $\mathbf{y}$ and $\mathbf{u}$ are 0.2477 and 0.1103, respectively. The sample size is $N = 1023$ and the system is excited with a full-band white PRBS signal.

Following DIPCA procedure, we guess a value for the number of constraints d, and check whether the last d eigenvalues are unity. If the last d eigenvalues exactly equal unity, the guessed d equals true number of constraints d^{*}. If not, we refine our guess of d to a higher or lower value. For the above example, eigenvalues obtained from guessed $d = 4$ gives

$$\Lambda = [28.3 \;\; 26.5 \;\; 15.6 \;\; 14.4 \;\; 11.2 \;\; 10.0 \;\; 9.3 \;\; 2.9 \;\; 1.0458 \;\; 1.0031 \;\; 0.9801 \;\; 0.9710]$$

Therefore, $d^{\star} = 4$. Order is given by $\eta = L - d^{\star} + 1 = 5 - 4 + 1 = 2$. Estimated noise variances are $[0.2460 \quad 0.1144]$. Monte-Carlo simulation of 200 runs for non-parametric estimates of confidence intervals yields:

$$y[k] + \underset{\pm(0.0453)}{0.4007}\, y[k-1] + \underset{\pm(0.025)}{0.5997} y[k-2] = \underset{\pm(0.0458)}{0.0012}\, u[k] + \underset{\pm(0.0654)}{1.2025}\, u[k-1] - \underset{\pm(0.0865)}{0.0013}\, u[k-2]$$

It is evident from the above example that DIPCA has correctly identified the process order, delay and the error-variances. Also, the parameter estimates obtained from DIPCA are unbiased.

5. Shortcomings of DIPCA

Although DIPCA in itself is a powerful algorithm that estimates all the unknown quantities, there are certain shortcomings in the current version of DIPCA which we study in the following subsections.

5.1. *Ambiguity in order determination for finite sample conditions*

For high order systems under finite sample conditions, it is observed that exact unity eigenvalues are not obtained even for a correct guess of the number of constraints d^{*}. In addition to this, the deviation in unity eigenvalues increases as the order η of the process increases, or the sample size N decreases.

For example, consider a 5th order process with unit delay, excited with a full-band white PRBS input signal as

$$\text{P2:} \quad y^{\star}[k] + 0.2y^{\star}[k-1] - 0.6y^{\star}[k-5] = 1.2u^{\star}[k-1] + 1.6u^{\star}[k-2]$$

The error-variances are such that the signal-to-noise ratio (SNR) is maintained at 10. Stacking lag L is arbitrarily chosen to be 15. The true number of linear relations will, therefore, be given by $d^* = L - \eta + 1 = 15 - 5 + 1 = 11$. However, for a guess of $d = 11$, DIPCA does not give exact unity eigenvalues for the smallest 11 eigenvalues, as depicted in the following plots:

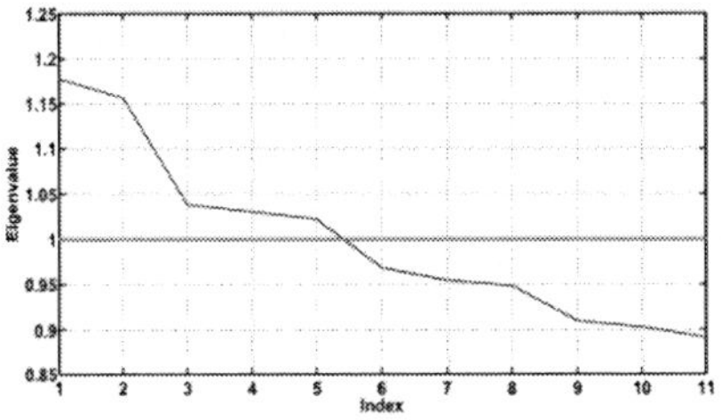

Figure 2: Last $d^* = 11$ λ's for $N = 1023$

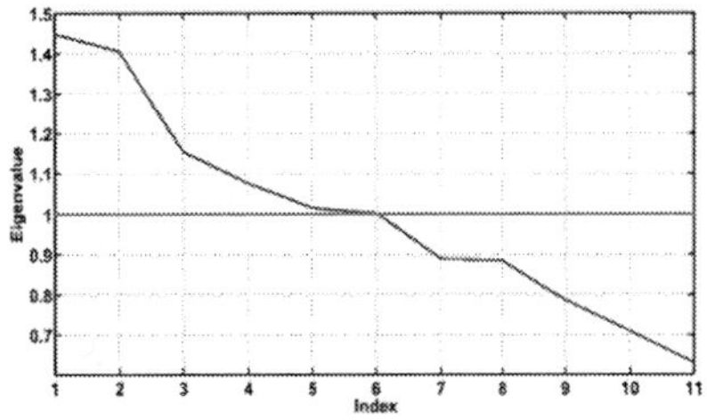

Figure 3: Last $d^* = 11$ λ's for $N = 127$

This deviation from unity leads to ambiguity in order determination since order determination is based on identifying the true number of linear relations d^* by counting the number of unity eigenvalues and then determining the order using the relation $\eta = L - d^* + 1$.

5.2. Effect of stacking in excess and existence of optimal stacking lag

Construction of the data matrix $\mathbf{Z}$ requires stacking lagged version of the input and output variables up to a lag L which is arbitrarily chosen to be large enough to accommodate higher order processes. However, there is a trade-off associated with choosing the value of L - higher L leads to higher redundancy, i.e. the number of equations that go into solving overdetermined set of equations for $\sigma^2_{e_u}$ and $\sigma^2_{e_y}$ increases which leads to a better least squares solution, but at the same time, the effective number of degrees of freedom $N - L - 2$ decreases leading increased variability of estimates.

To demonstrate this trade-off, again consider the process P2. Σ_e is estimated at different stacking lags, L and the mean absolute error (MAE),

$$\text{MAE} = ||\text{diag}(\Sigma_e^0 - \hat{\Sigma}_e)||_1/2 \tag{8}$$

is computed for different L. For a given L, MAE is averaged over 100 different realizations of error-sequences maintaining the SNR at 10. The following plots for MAE vs. L are obtained:

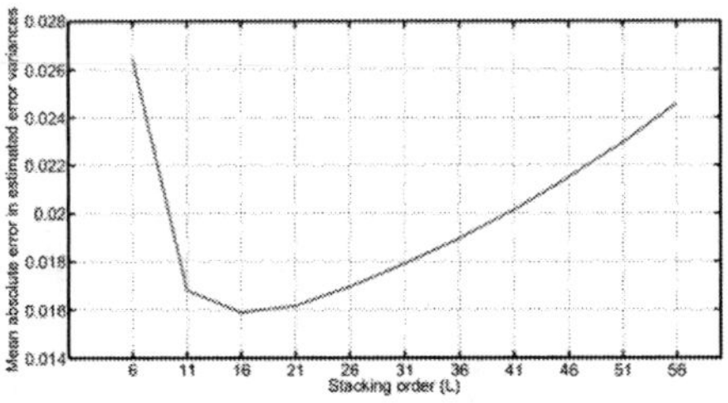

Figure 4: $N = 500$

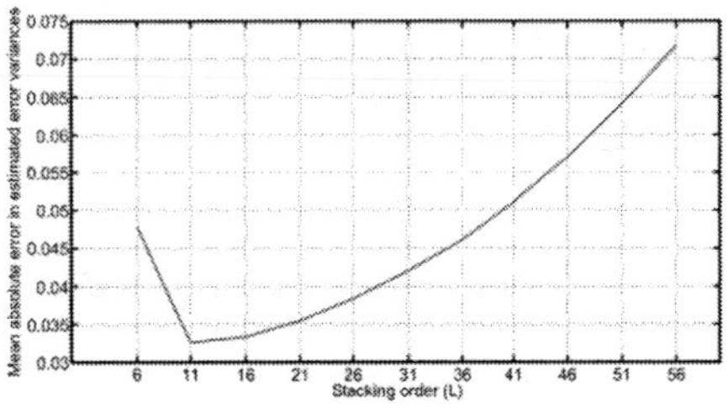

Figure 5: $N = 200$

$\hat{\Sigma}_e$ for $N = 500$ and $L = 16$ is $[0.2201 \quad 0.0957]$ when the true values are $[0.2322 \quad 0.0937]$. $\hat{\Sigma}_e$ for $N = 200$ and $L = 11$ is $[0.2280 \quad 0.0816]$ when the true values are $[0.2320 \quad 0.0952]$. Note how optimal L changes with the sample size - the optimal L decreases with decrease in N and increases with increase in N.

6. Proposed metric for resolving order ambiguity

The deviations in eigenvalues are observed since the results of IPCA are derived under asymptotic conditions i.e. as $N \to \infty$. As the sample size decreases, this conditions is invalidated and the estimates of eigenvalues are distorted. The distortion happens in both the directions i.e. some of the eigenvalues are over-estimated and some are under-estimated. It is, however, observed that the net deviation in the smallest d eigenvalues equals 0 until the point when $d \leq d^*$. Building upon these properties, we propose a technique to resolve this issue.

6.1. 'd-SEVR' procedure for order determination:

We propose the following systematic procedure to identify the true number of constraints d^* and to subsequently identify the true order η from diverging eigenvalues.

Table 1: Proposed 'd-selective eigenvalue ratio procedure' for order determination

1. For given guessed number of constraints d, calculate the ratio of the maximum (λ^{max}) and minimum (λ^{min}) eigenvalues among the last d eigenvalues, i.e.

$$R_d = \lambda_d^{max}/\lambda_d^{min}$$

2. Calculate R_d for different guesses of d, preferably increasing in steps of 1.

3. The value of d after which there is an abrupt jump (inflection point) in the magnitude of R_d gives us the true number of constraints d^*.

4. The order is given by $\eta^* = L - d^* + 1$.

6.2. Illustrating example

For the same 5th order process P2, we employ the *d-SEVR* procedure to identify d^* for datasets with sample sizes $N = 1023$ (Fig. 6) and $N = 127$ (Fig. 7).

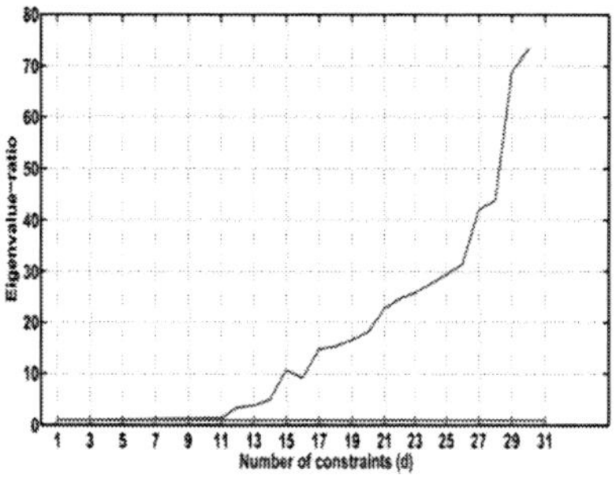

Figure 6: R_d vs. d for $N = 1023$

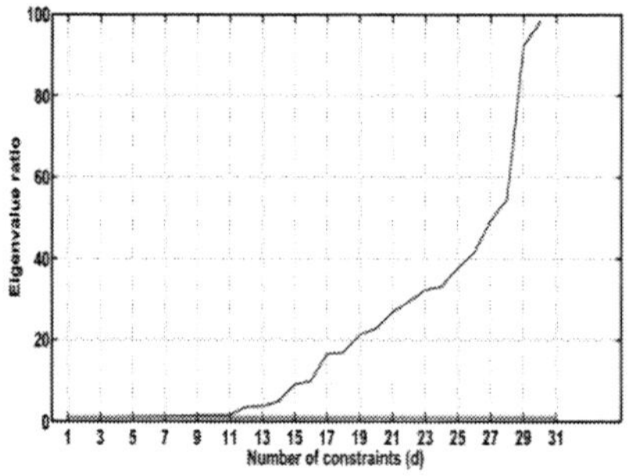

Figure 7: R_d vs. d for $N = 127$

From the above simulation results, it is clear that the proposed metric R_d is robust towards the number of samples N and at the same time is sensitive to an overestimated guess of d.

7. Applications in data reconciliation

A major end goal of any model building exercise is to generate noise free estimates of the noisy data coming out of sensor measurements. In this section, we discuss the application of DIPCA in data reconciliation from a practical viewpoint.

7.1. PCA for dynamic data reconciliation

Since DIPCA is based on the PCA framework, one could consider generating noise-free estimates of the input-output variables using PCA de-noising technique. However, this results in non-unique estimates even for the same variable because in our data matrix,

$$\mathbf{Z} = \begin{bmatrix} y[k] & y[k-1] & u[k-L] \\ y[k+1] & y[k] & u[k-1+1] \\ \vdots & \vdots & \vdots \\ y[N-L] & y[N-L+1] & u[N] \end{bmatrix}$$

we have stacked lagged versions of the same input and output variables. PCA considers all these $2 \times (L+1)$ variables as separate variables independent from each other. As a result of this, we get $L+1$ different reconciled estimates each of the two variables, $\mathbf{y}$ and $\mathbf{u}$.

The following plot corresponding to two different estimates of $\hat{\mathbf{y}}$ for the process P2 depicts this:

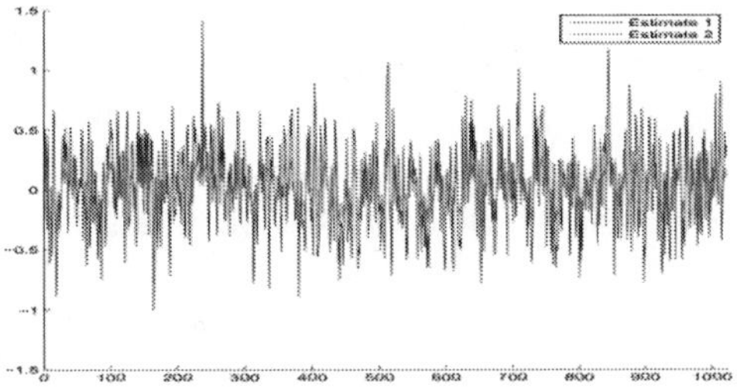

Figure 8: Estimated (filtered) values $\hat{\mathbf{y}}$ from first two columns in $\hat{Z}$

Therefore, we need a more rigorous and technically correct way of generating reconciled values of the input-output variables, and hence, we turn to a Kalman Filter based approach to solve our data reconciliation problem.

7.2. EIV Kalman Filter formulation

In order to get unique, noise free estimates of the input and output variables, we switch to a Kalman Filter based approach since for a linear process, Kalman Filter generates noise free estimates in an optimal manner. However, since the problem setting is that in the errors-in-variables (EIV) framework, we cannot use the usual Kalman Filter as it is. Instead, we use the EIV Kalman Filter proposed in [8] which is formulated as follows:
Consider the state-space form of any process given by

$$\begin{aligned} x[k+1] &= Ax[k] + Bu^*[k] + Gw[k] \\ y^*[k] &= Cx[k] + Du^*[k] \end{aligned} \tag{9}$$

where, $u[k] = u^*[k] + e_1[k]$ and $y[k] = y^*[k] + e_2[k]$ are the input and output variables with measurement errors. In terms of measurements $u[k]$ and $y[k]$, the above model can be re-written as

$$\begin{aligned} x[k+1] &= Ax[k] + Bu[k] - Be_1[k] + Gw[k] \\ y[k] &= Cx[k] + Du[k] - De_2[k] + e_2[k] \end{aligned} \tag{10}$$

Introducing auxiliary white noise processes,

$$\eta_x[k] = Gw[k] - Be_1[k] \tag{11}$$

$$\eta_y[k] = e_2[k] - De_1[k] \tag{12}$$

the above model can again be re-written as

$$\begin{aligned} x[k+1] &= Ax[k] + Bu[k] + \eta_x[k] \\ z[k] &= Cx[k] + \eta_y[k] \end{aligned} \tag{13}$$

where,

$$z[k] = y[k] - Du[k] \tag{14}$$

From the equations given above, the update rules for state estimation can be derived by minimizing the trace of the estimated states' error-covariance matrix as given in [8].

Table 2: Proposed Methodology

1. Convert the estimated difference equation model into its equivalent state-space form.
2. Estimate unknown states of the state-space model using EIV Kalman Filter update rules.
3. Compute the filtered $\mathbf{y}$ and $\mathbf{u}$ from the estimated states using equations (13) and (14).

7.4. Illustrating example

Consider the data generating process,

$$\text{P3:} \quad y^*[k] + 0.4y^*[k-1] + 0.6y^*[k-2] = 1.2u^*[k-1] + 1.6u^*[k-2]$$

The error variances are such that the SNR is 10 with error variances $\sigma^2_{e_u} = 0.0930$ & $\sigma^2_{e_y} = 0.2336$. Model identified using DIPCA with a stacking lag $L = 5$, averaged over 10 different noise realizations is

$$\bar{\mathbf{A}} = [1 \;\; 0.3955 \;\; 0.5938 \;\; 0.0030 \;\; -1.2035 \;\; -1.5867]$$

or,
$$y(k) + 0.3955y(k-1) + 0.5938y(k-1) = 1.2035u(k-1) + 1.5867y(k-2)$$

The estimated error-variances are $\sigma^2_{e_u} = 0.0881$ & $\sigma^2_{e_y} = 0.2436$. Using *tf2ss()* in MATLAB, the equivalent state space representation of the above model is obtained as

$$\begin{aligned} x[k+1] &= \begin{bmatrix} -0.3955 & -0.5938 \\ 1.000 & 0 \end{bmatrix} x[k] + \begin{bmatrix} 1 \\ 0 \end{bmatrix} u[k] \\ z[k] &= \begin{bmatrix} 1.2035 & 1.5867 \end{bmatrix} x[k] \end{aligned}$$

The state estimation error-covariance matrix for this model is $\mathbf{P} = \begin{bmatrix} 0.1060 & -0.1027 \\ -0.1027 & 0.0855 \end{bmatrix}$. The reconciled values for the output variables are obtained and plotted against their measured values below in Figure 9. Also, since the goodness of fit gives us an idea of the amount of de-noising that has happened, Table 3 shows the goodness of fit between the noise free output variable y^* and the measured y & filtered values $\hat{y}$. We can see that there is a 3% increase in the goodness of fit for $\hat{y}$ generated using EIV Kalman Filter.

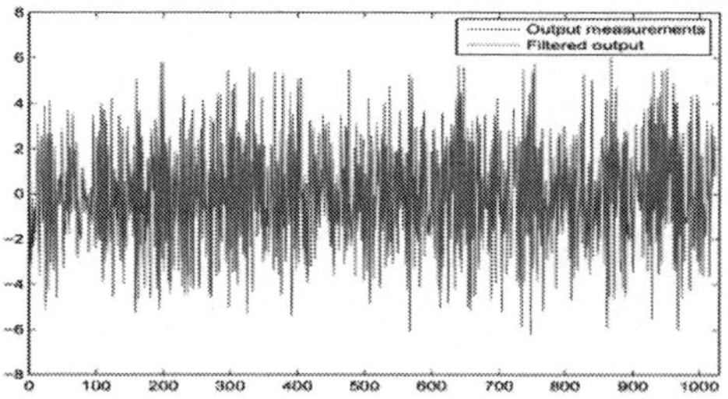

Figure 9: Measurement and reconciled output variables

Table 3: Goodness of fit NMSE values

Goodness of Fit	
Between $\mathbf{y}^*$ and $\mathbf{y}$	0.9348
Between $\mathbf{y}^*$ and $\hat{\mathbf{y}}$	0.9639

8. Conclusions

In this paper, DIPCA algorithm is studied from a practical viewpoint and its shortcomings under small sample conditions are highlighted. A new metric for determining the model order is proposed which, through simulations, is shown to be robust towards the sample size and sensitive to the true model order. Existence of an optimal stacking lag is also shown and its determination from data is the problem that we will address in the future. Kalman Filtering for the errors-in-variables setting is used to generate reconciled values of the input and output variables in an optimal manner. Since multiple input multiple ouput (MIMO) systems also have huge prevalence in industry, extension of DIPCA algorithm for identification of such systems will be our future direction of research.

References

[1] Soderstrom, T. (2007). "Errors-in-variables methods in system identification". *Automatica*, 43(6), 939-958.

[2] Ku, W., Storer, R., and Georgakis, C. (1995). "Disturbance detection and isolation by dynamic principal component analysis". *Chemometrics and Intelligent Laboratory Systems*, 30, 179-196.

[3] Guerfel, M., Othman, K. B., & Benrejeb, M. (2009). "On the structure determination of a dynamic PCA model using sensitivity of fault detection". IFAC Proceedings Volumes, 42(11), 958-963.

[4] Rato, T. J., & Reis, M. S. (2013). "Defining the structure of DPCA models and its impact on process monitoring and prediction activities". Chemometrics and Intelligent Laboratory Systems, 125, 74-86.

[5] Vijaysai, P., Gudi, R., and Lakshminarayanan, S. (2005). "Errors-in-variables-based modeling using augmented principal components". *Industrial & Engineering Chemistry Research*, 44, 368-380.

[6] Narasimhan, S., & Shah, S. L. (2008). "Model identification and error covariance matrix estimation from noisy data using PCA". Control Engineering Practice, 16(1), 146-155.

[7] Maurya, D., Tangirala, A. K., & Narasimhan, S. (2016). "Identification of Linear Dynamic Systems using Dynamic Iterative Principal Component Analysis". IFAC-PapersOnLine, 49(7), 1014-1019.

[8] Diversi, R., Guidorzi, R., & Soverini, U. (2005). "Kalman filtering in extended noise environments". IEEE Transactions on Automatic Control, 50(9), 1396-1402.

HIGH PRESSURE RATIO COMPRESSOR (HPRC) PROTOTYPE SURGE EXPLORATION TESTS AND SECOND QUADRANT CHARACTERISTIC DYNAMIC MODELING

Fabio Baldanzini, GE Oil & Gas, Florence, Italy
Mirco Calosi, GE Oil & Gas, Florence, Italy
Marco Pelella, GE Oil & Gas, Florence, Italy

Abstract

High Pressure Ratio Compressor (HPRC) technology is based on an innovative architecture that combines unshrouded and shrouded impellers on a single high-speed shaft to achieve pressure ratios and efficiency levels higher than other available technologies. As final step of the product development and validation, a test campaign has been carried out aimed to test the performance and explore the operability of the new machine.

Surge exploration tests have been also performed allowing an evaluation of the transient behavior and the mechanical robustness of the compressor even during a critical event such as Surge.

Compressor Surge has been analyzed under different conditions, forcing the operating point to move beyond the performance envelope in full speed, at low and high pressure levels, and during emergency shut-downs. Thanks to the complex arrangement of the gas loops and several valves used to recycle back the compressed gas, different level of Surge intensity were induced upon emergency shut-down.

The explorations of a fast event like Surge called for special instrumentations, configured to acquire process data both in direct and reverse flow condition in the most reliable way, with a high frequency sampling for an oil and gas environment.

The result of this work is a breakthrough for the tuning of a centrifugal compressor model – not only HPRC ones – to be used for dynamic simulations and prediction of compressor dynamics during Surge events (like Surge cycle frequency and absorbed torque during Second Quadrant operation) in more reliable and robust way.

Surge exploration tests results analysis in terms of vibrations and thrust loads, together with development of a compressor enhanced dynamic model, allowed to move from a Surge acceptance criterion - based on the time spent on the left of the Surge Limit Line during an Emergency Shutdown event - to a more physics based criterion - based on the acceptable number of Surge cycles, thus increasing selection optimization of additional protections like hot/cold gas bypass valves.

1. Introduction

HPRC compressor prototype has been developed as a combination of stacked unshrouded and shrouded impellers on a single high-speed shaft to reduce the number of stages up to a factor 2 with respect to a traditional compressor. HPRC allows pressure ratios and efficiency levels higher than other available technologies, reducing compressor string footprint and weight.

A test campaign on the HPRC prototype has been carried out to evaluate compressor performance, aeromechanical and rotordynamic behavior both in steady state and upon the most critical scenario like compressor Emergency Shut Down (ESD) and Start-Up [1].

The HPRC prototype is a 3 sections compressor arranged in line with two, three and two impellers respectively. The two first section impellers are unshrouded type.

As final step of the product development and validation, a test campaign has been carried out to test the behavior of the compressor prototype beyond the First Quadrant envelope, focusing on the behavior at Surge. This activity allowed evaluating the mechanical robustness of the compressor upon a critical event such as deep Surge conditions (Second Quadrant Operation).

The purpose of Surge Exploration Tests was also to collect thermodynamic parameters beyond the First Quadrant compressor envelope. In particular, Second Quadrant operation and periodic Surge cycles in which the compressor experiences a series of flow reversal and recovery, was deeply investigated in order to develop a reliable model - capable to predict compressor behavior upon Surge.

The paper is organized as follows: in section 2, the HPRC test bench gas loop and the instrumentation used are presented; section 3 describes the HPRC test campaign; in section 4 vibrations, axial displacements and thrust loads recorded during two runs of the HPRC test campaign are analyzed; section 5.1 describes the results of a Model Test performed on a single stage compressor, used to characterize the second quadrant map branch and how it has been adapted to the HPRC machine; in section 5.2 overall gas loop dynamic simulation model is presented; section 5.3 deals with data match between simulation results and test data; finally, conclusions are given in section 6.

Figure 1. HPRC Test bed arrangement at the GE Oil & Gas Facility in Florence (Italy)

2. Gas loop arrangement and instrumentation

The testing loop arrangement consisted of three independent closed gas loops laying on each HPRC section, equipped with recycle solenoid valves (ASVs) and cooler at the suction. Bypass valves (PCV12 and PCV23, ball valve type) were installed to reproduce serial operation. Such configuration (Figure 2) provided the maximum flexibility to test the prototype in several operational arrangements, allowing also decoupling the three phases as well as study their mutual interactions occurring over a Surge event.

Compressor prototype was coupled with a variable speed electric motor driver with 10 MW maximum available power.

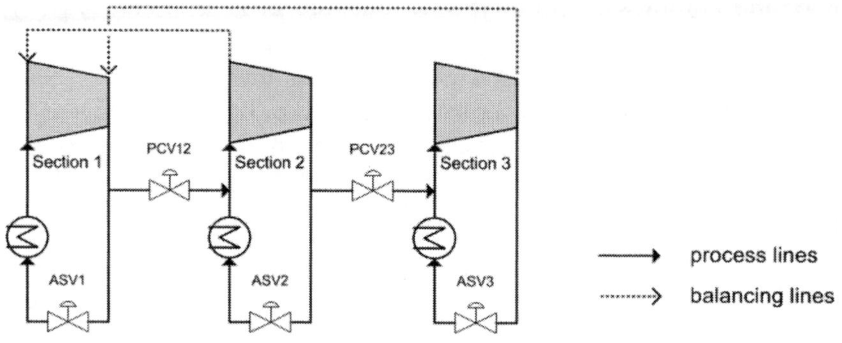

Figure 2. Testing loop arrangement

To evaluate Surge both from the thermodynamic and mechanical standpoint, the test bed was equipped with dedicated instrumentation aimed to properly collect data in direct and reverse flow operation conditions, also with suitable acquisition frequency for Surge phenomena (10Hz for flows and pressures and 1Hz for temperatures). Four dynamic pressure transducers and four thermocouples were installed on suction and discharge of each compressor section; flow rate measurement was performed by differential pressure transducers installed both on suction and discharge, with 2% of accuracy on design values. The orifice on each discharge was installed reversed to acquire reverse flow across the HPRC sections. Radial vibrations were measured through Bently Nevada proximity probes located on both shaft ends with a 2000 Hz frequency response bandwidth. Axial displacements and axial thrust acting on the rotor were respectively measured through Bently Nevada proximity probes and load cells installed on the thrust bearing.

3. Surge exploration tests campaign

Compressor Surge has been analyzed under different operating conditions, forcing the operating point to move beyond the performance envelope at full speed and during emergency shut-downs, at low and medium gas loop pressure. Compressor control algorithm was modified to induce different levels of Surge intensity upon Emergency Shut Down through an instantaneous (as per standard procedure), delayed and frozen recycle valves opening action.

Compressor Surge at full speed was induced on a specific compressor section slowly closing the relevant ASV to increase the pressure ratio and reduce the flow until the operating point move beyond compressor envelope Surge limit. In order to prevent any mutual effects upon Surge, the two other compressor sections were kept operating far from Surge limit, in the middle of the compressor envelope, close to the design point. In this way, Surge cycles on each compressor section were evaluated.

Before and after each Surge test, performance at a reference point condition was evaluated in order to capture any potential degradation due to Surge phenomena detrimental effects on the machine. Compressor performance at reference point was always met again demonstrating that no detrimental effect due to Surge on the compressor mechanical integrity was experienced.

The present work refers to Medium Pressure Surge Tests, being the most meaningful test runs in terms of Surge severity.

A compressor mechanical analysis upon Surge event has been performed leveraging on data relevant to test performed on section 1 at full speed and on data relevant to ESD test with all recycle valves kept at their initial opening position.

Data match with the dynamic model results has been then executed on the former two tests data and on data relevant to tests performed on section 2 and 3 at full speed.

4. Mechanical test data analysis

Mechanical behavior has been analyzed in terms of radial vibrations, axial displacements and thrust load variations. Figure 3a reports compressor speed and radial vibrations evaluated through proximity probes located on NDE compressor side. The red line depicts the direct radial vibration, including all frequency spectrum contributions; instead, the green one is the radial vibration at the revolution frequency (hereafter called in this study "1xRev"). Both trends have been normalized with respect to the value of the direct radial vibration at the initial steady state condition. To be noted that direct radial vibration is reported as Peak-Peak, while 1xRev is 0-Peak. In the initial steady state, these contributions were identical, as indication that radial vibrations were mainly due to rotor residual unbalance. On the contrary, during Surge, non-synchronous vibrations were excited so that direct radial exceeded 1xRev vibrations: according to Figure 3a, the former reached about 2.2 and the latter 1.8 (normalized values). Therefore, Surge cycles were such to excite sub-synchronous vibrations for an amount close to 0.4. To be noted that maximum allowable threshold (1.7) was overcome without triggering trip. In fact, in order to disregard spurious data, a signal continuously above the threshold for at least 0.3 sec was required to initiate the ESD.

To detect the main contribution frequency to vibrations spectrum during Surge, Waterfall and the spectrum plots were also developed for radial vibrations probes. As done for the other variables, vibration frequency has been normalized in the following figures with respect to the value at initial steady state condition. Figure 3b highlight that rotor Surge response was evident in the sub-synchronous range of 1xRev frequency and only during the flow inversion (from positive to negative) timeframe. A forced response on first rotor mode is present within this range (at about 0.25) corresponding to the maximum spectrum contribution (red box in Figure 3b).

Also, the compressor speed variation was deemed a key parameter for compressor behavior upon Surge as it is an indication of the absorbed torque variation during the transient. As the operating point moved in the Second Quadrant operation, speed increased as effect of a lower absorbed torque, reaching a maximum value while approaching the zero flow condition. Then, when operating point stepped back into First Quadrant, the absorbed torque started increasing leading to a speed drop till its minimum value. To be noted that speed variations were also affected by speed controller, which was acting on electric motor torque trying to keep the speed at its set point. However, this effect has to be considered negligible on speed variation with respect to the compressor absorbed torque fluctuations due to Surge.

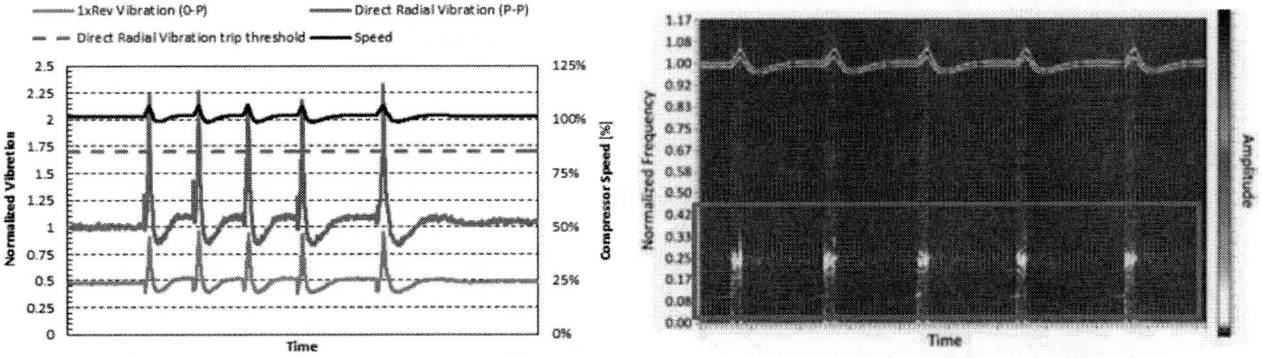

a) Radial Direct and 1xRev vibration trends and compressor speed

b) Waterfall for Radial Direct Vibration

Figure 3. First Section Surge Cycles at full speed (radial vibration probe on NDE)

The compressor Surge behavior was also tested upon ESD, forcing all three sections to dash in reverse flow condition (Second Quadrant) by freezing the ASVs opening position so to hinder the suction and discharge pressure equalization during coast down.

Before initiating ESD, the three compressor Sections were brought close to the SLL by closing ASVs. As soon as the ESD was triggered, speed started decreasing and reverse flows were detected in all three sections. As soon as reverse flow condition across the compressor occurred, direct radial vibrations showed a peak amplitude of 1.5 (126% of initial operation) as shown in Figure 4a. The same increment was not present on the 1xRev component trend, as indication of a wider content of frequencies was contributing to the overall vibration upon Surge.

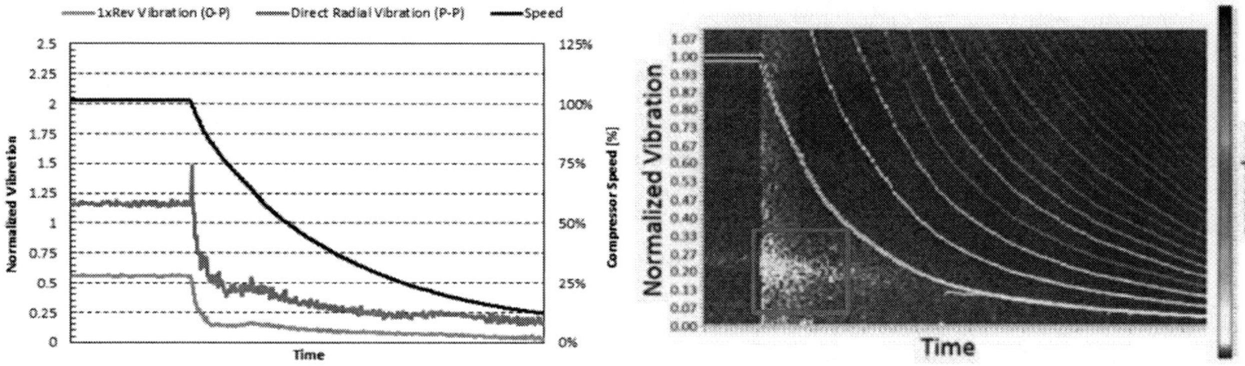

a) Radial Direct and 1xRev vibration trends and compressor speed

b) Waterfall for Radial Direct Vibration

Figure 4. Surge upon ESD (radial vibration probe on NDE)

Waterfall plot in Figure 4b provides a further confirmation that Surge is a "broadband" transient phenomenon. Vibrations amplitudes decayed without any further peaks during the compressor coast down, as indication that reverse flow condition is an acceptable operation from a mechanical standpoint. As regards the entire vibrations spectrum, the same conclusions relevant to the Surge tests at full speed can be drawn since sub-synchronous vibrations of a minor magnitude were observed (red box in Figure 4b).

For the sake of completeness, axial displacements and axial thrust load were also evaluated, as done in [7]. Figure 5 shows these variables normalized with respect to the value at their own initial steady state condition. In case of Surge under operating conditions, axial displacements and thrust load had a peak to 1.24 and 2.5 respectively (Figure 5a), without exceeding the thrust bearing rated capacity.

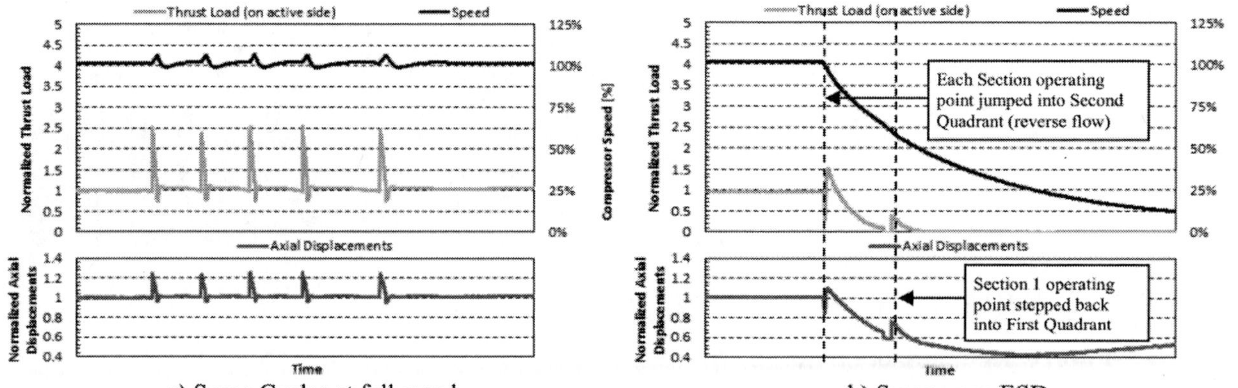

a) Surge Cycles at full speed

b) Surge upon ESD

Figure 5. Axial Displacements and Thrust Load on active side

On Surge upon ESD test, both axial displacements and thrust load decayed during the compressor coast down apart two minor increments experienced as soon as the operating point moved in the Second Quadrant region and when first section compressor stepped back from Second to First Quadrant (see Figure 5b).

An important outcome of Surge exploration tests is that a reverse flow condition (aka Second Quadrant region) can be withstood by compressor even for a prolonged operation since vibrations and axial thrust trends decayed smoothly. However, being Surge phenomenon associated to forward and backward flow, not only across the compressor but even to compression system equipment (like scrubbers, coolers, valves, piping and relevant anchors, etc.), a threshold on the number of Surge cycles shall be considered in order to avoid undesirable upsets due to pressure waves and vibrations.

A new Surge acceptance criterion has been developed accounting for a maximum of three Surge cycles instead of operating time beyond the compressor Surge Limit Line (SLL) as per the current commonly used design practice.

The new approach to evaluate compressor Surge is still conservative, but at the same time, prevents undue complexity such as installation of Surge protection additional devices (like HGBV or CGBV) or hardware modifications.

5. Surge model and data match

The purpose of Surge Exploration Tests was also to collect thermodynamic data relevant to the Second Quadrant operation to be compared with simulation results.

For this scope, compressor behavior has been simulated with a custom model based on a modified Moore-Greitzer [5] compressor dimensionless characteristic curve (cubic shape). The Second Quadrant map branch has been characterized with an ad-hoc Model Test carried out on a single stage compressor [3]. The model has been then adapted on the HPRC compressor based on Surge Exploration Tests, subject of this paper. Compressor map development in the second quadrant and HPRC Surge tests data match are described in the following sections.

5.1. Second Quadrant compressor model

In the Model Test arrangement (which is more deeply investigated in [3] and [4]) a booster compressor was connected in series with the tested stage forcing the flow to be stable in reverse flow condition (Figure 6).

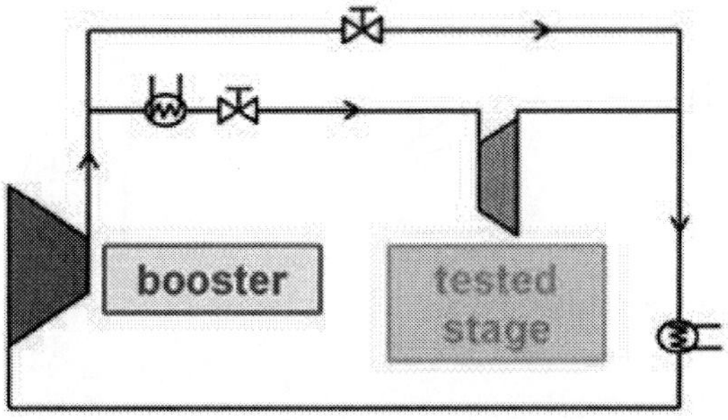

Figure 6. Model Test Arrangement

The trends of pressure ratio measured in the second quadrant are reported as a function of flow coefficient in Figure 7.

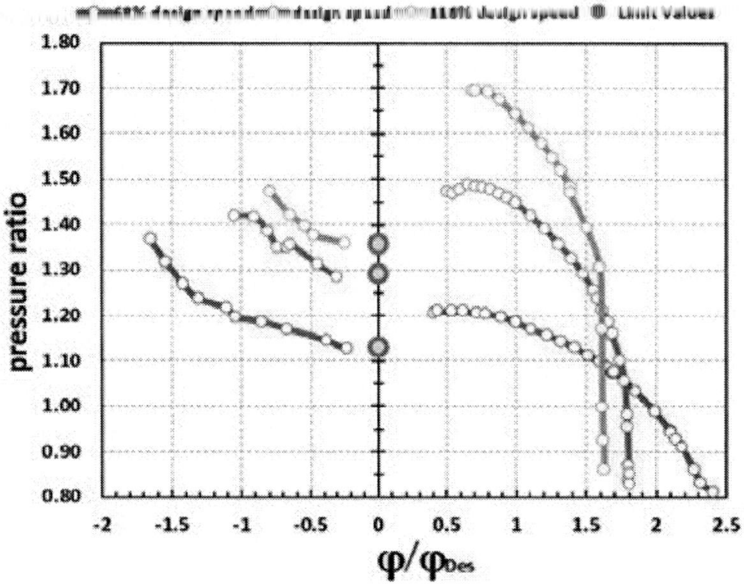

Figure 7. Measured Performance Second Quadrant

$$\varphi = \frac{4\dot{m}}{\rho_{in}\pi D_2^2 u_2} \qquad \textbf{Eq. 1}$$

In the first quadrant (positive flow), compressor curves have the well-known shape: decreasing the flow, the pressure ratio that the compressor is able to deliver increases till reaching the surge limit. In the second quadrant (negative flow), reducing the amount of reverse flow to zero, pressure ratio tends to a minimum limit value which corresponds to centrifugal forces at the imposed rotational speed.

Compressor curves trends observed are in accordance with the cubic shaped compressor dimensionless characteristic curve of Moore-Greitzer model (Figure 8) which has been taken, together with the equation of motion (Eq. 2) as reference for the development of the compressor dynamic model.

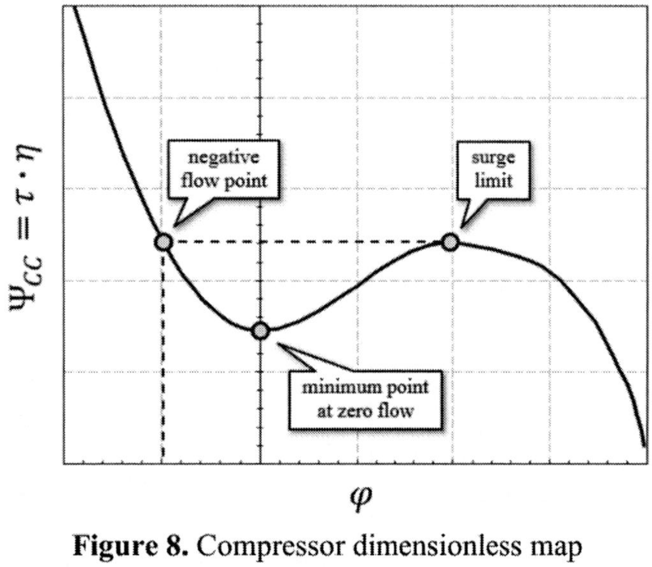

Figure 8. Compressor dimensionless map

$$\begin{cases} \eta_{pol} = f(\varphi) \\ \Psi_{CC} = \dfrac{H_{pol,CC}}{u_2^2} = f(\varphi) \\ \mathrm{H}_{pol,P} = f(T_{in}, P_{in}, P_{out}) \\ \dfrac{\partial \varphi}{\partial t} \propto \left(\mathrm{H}_{pol,CC} - H_{pol,P}\right) \end{cases} \qquad \textbf{Eq. 2}$$

In Eq. 2, $H_{pol,P}$ is the polytropic head calculated from the process gas thermodynamic properties at the inlet and outlet section of the compressor, $\mathrm{H}_{pol,CC}$ is the polytropic head of the compressor given by the compressor map and η_{pol} is the compressor polytropic efficiency also given by the compressor map (for more details on equation of motion refer to [5]).

The three stage curves of the HPRC compressor has been tuned matching dynamic simulation results with maximum and minimum direct/reverse flows measured during the Surge tests. Fine tuning parameters used are the coordinates of minimum point of the dimensionless polytropic head Ψ_{CC} at zero flow and coordinates of negative flow point at which the dimensionless polytropic head is equal to the one at surge limit (see Figure 8).

5.2. Dynamic Simulation Model

With reference to the actual gas loop (Figure 2), a HPRC compressor loop dynamic model (Figure 9) has been built with a detailed characterization of all the testing equipment like valves, piping, coolers, etc.

The software used was Aspen HYSYS® configured with the Peng-Robinson equation-of-state and Lee-Kesler option for the calculation of enthalpies. Standard Aspen HYSYS® compressor element has been replaced by a custom element able to correctly simulate the behavior of the compressor in the Second Quadrant according to the model described in the previous section.

Compressor speed model used is given by a torque balance equation (Eq. 3) applied to the train, in which inertia of all the elements composing the train (electric motor driver, gear box, couplings and compressor) and the speed controller acting on electric motor torque have been considered. Such model allowed simulating also the speed fluctuation due to the variations of both absorbed torque by the compressor and available torque from the electric motor.

$$\frac{\partial \omega}{\partial t} = \frac{\frac{T_{EM}}{GR} - T_{CC} - f \cdot \omega}{I_{TOT}} \qquad \textbf{Eq. 3}$$

In the dynamic model, the speed controller has been tuned so to have its action on Electric Motor Torque (T_{EM}) reproducing the speed variation experienced during full speed tests. However, it has to be noted that the main parameter affecting the speed variation in Eq. 3 is the compressor absorbed torque (T_{CC}), which is calculated by the custom compressor element using the aforementioned curves.

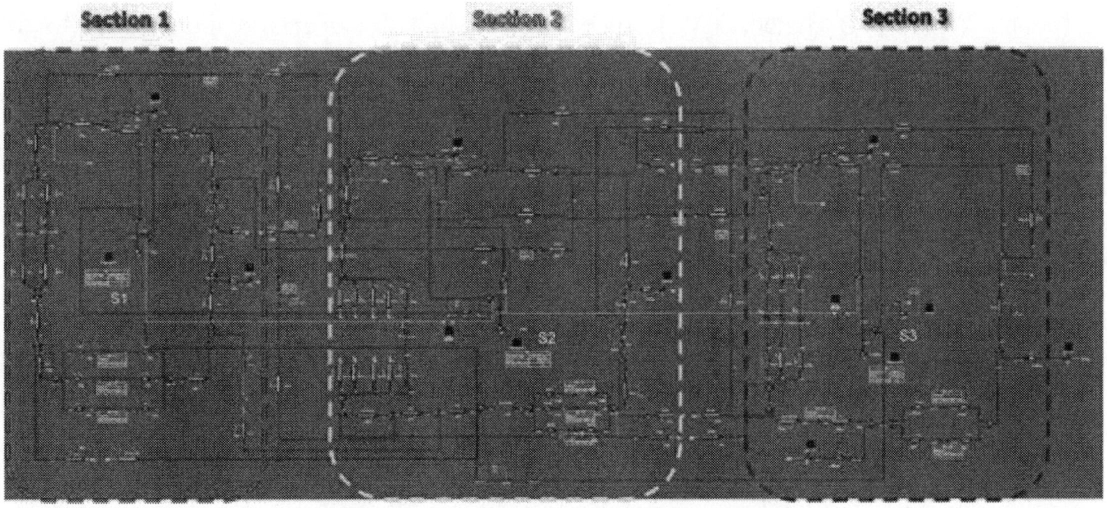

Figure 9. Dynamic Simulation Model

The HPRC compressor loop dynamic model has been first validated matching initial steady state condition of each test run in terms of suction and discharge pressure, temperature, mass and volumetric flow of each compressor stage, and then used to replicate data obtained in the tests where Surge were induced on compressor section 1, 2 and 3 (separately) at full speed and upon ESD with all recycle valves kept at their initial opening position. Data matching results are presented in the next section.

5.3. Simulation data match

Data matching study results are shown in Figure 10 and Figure 11 for the four runs analyzed. Variables under investigations in this study are: mass flows, suction and discharge pressure in each compressor section and train speed. Compressor suction and discharge temperatures have not been included in this analysis as temperature probe response time and thermal inertia of the system does not allow to catch properly the dynamic of a such fast event like surge.

Even if all the variables mutually influence each other and so it is not possible to analyze them individually, data match of a single variable can give an indication of the accuracy of a particular part of the dynamic model.

The good data match of pressures in both, full speed and emergency shut-down tests, is mainly indication that valves and all volumes of the gas loops have been simulated correctly.

Train speed predictability indicates mainly the capability to model the compressor speed control system and accuracy of torque balance equation. Model accuracy in predicting train speed can be better assessed in the full speed tests, where compressor speed control system is active and modulating electric motor delivered torque. In order to achieve a good speed matching, an accurate estimation of compressor absorbed torque is necessary, especially in the Second Quadrant. In the First Quadrant, the Torque can be calculated as follow:

$$T_{CC} = \frac{\frac{\mathrm{H}_{pol,CC}}{\eta_{pol}} \cdot \dot{m}}{\omega} \qquad \textbf{Eq. 4}$$

In the Second Quadrant, Eq. 4 can be considered to be still valid: as soon as the operative point dash to the Second Quadrant, the absolute value of polytropic head and efficiency does not vary significantly, while the mass flow decrease till almost 10-20% of the direct flow condition (see both Figure 10 and Figure 11). Hence, in full speed tests, when the operative point dash to the Second Quadrant, the compressor absorbed torque decreases significantly leading, according to torque balance equation (Eq. 3), to an increase of the compressor speed. This behavior has been correctly caught by dynamic simulation, as shown in Figure 10 (black lines).

Finally, predictability of mass flow processed by each compressor section is an indication of the accuracy of the model of compressor itself.

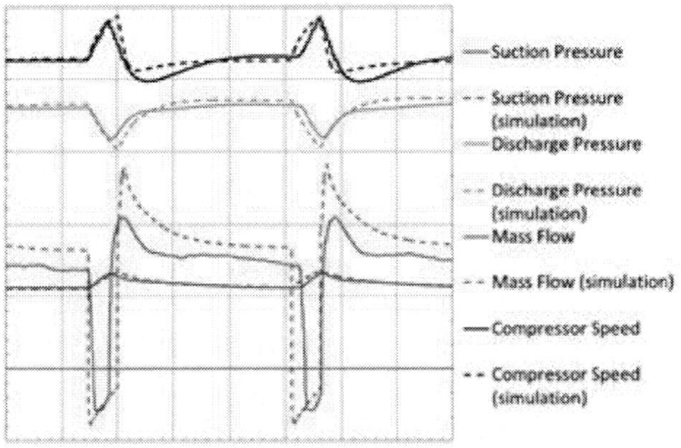

a) Section 1 Surge at full speed

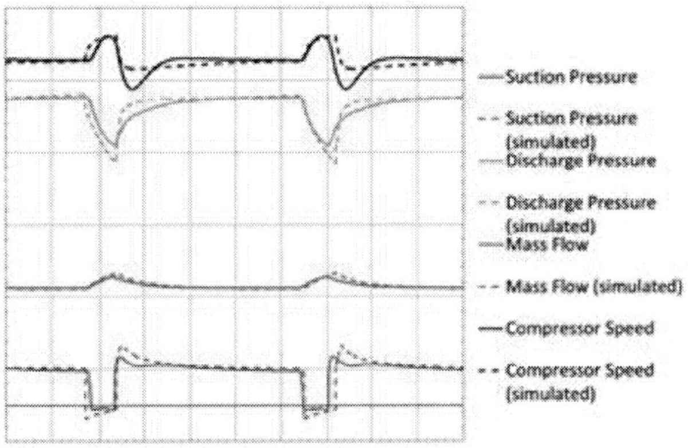

b) Section 2 Surge at full speed

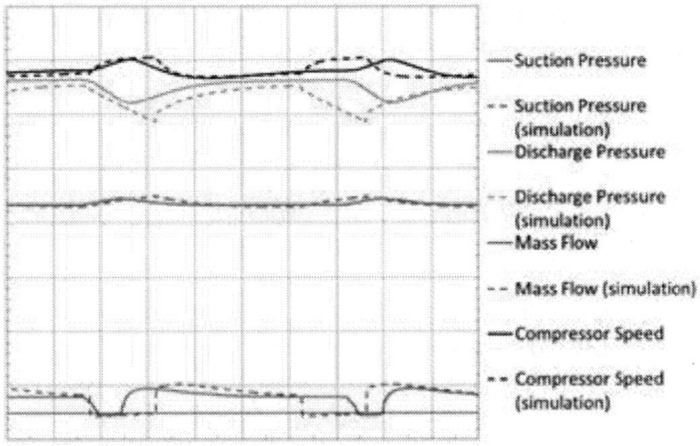

c) Section 3 Surge at full speed

Figure 10. Surge tests data match between dynamic simulation and test data (full speed)

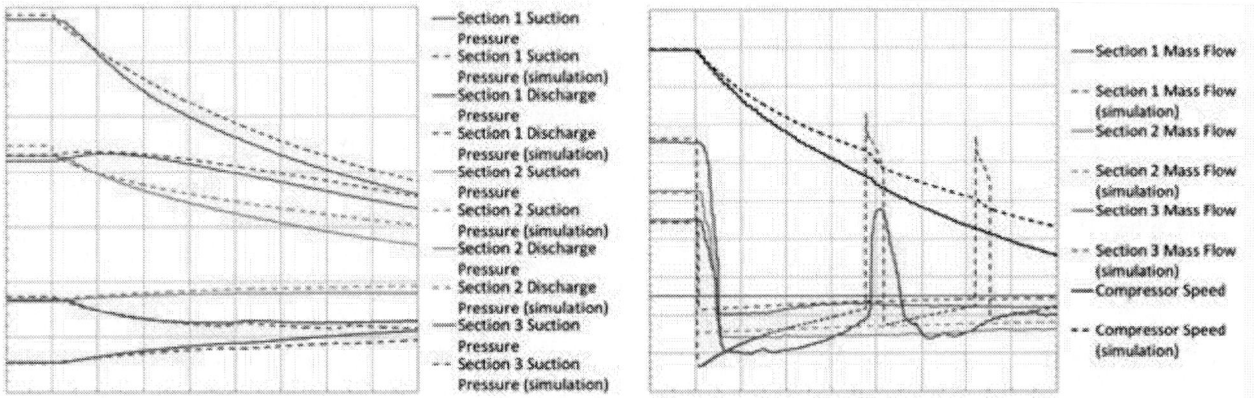

Figure 11. Surge tests data match between dynamic simulation and test data – ESD with Recycle Valves frozen

As highlighted above, when the operative point is recovering from a reverse flow condition, i.e. flow processed by the compressor is increasing, the train experienced also a speed increase. Since speed has a direct influence on the compressor deliverable head (compressor deliverable head is proportional to the square of the speed), also equation of motion (Eq. 2) is affected: the more the compressor deliverable head, the higher the derivative of flow and so the less time is spent in the second quadrant.

Derivative of flow, and so the time spent in the second quadrant, is also affected by the process polytropic head and so by the rate of suction/discharge pressures equalization which is strongly dependent on plant volumes, reverse mass flow through the compressor and mass flow passing through the recycle valves. This lead to the conclusion that all the variables of the compressor loop have an influence on Surge cycles parameters: reverse flow duration, number of cycles and maximum and minimum direct/reverse flows. Since these are the key parameters on which the Surge assessment of the machine during ESD is based, a good dynamic simulation model, capable to predict correctly all the variables that describes the compressor loop behavior during both normal running and ESD scenarios, is essential to properly design Surge protection devices.

Looking at tests results data match from a Surge cycles parameters standpoint, it can be observed that:

- as described in the previous section, maximum and minimum direct/reverse flows were the parameters fitted in the tuning process of the three compressor stage curves. Indeed, the simulated values fit almost perfectly the field data;
- time spent in reverse flow has been fairly matched in section 1 and 2 surge at full speed and ESD tests (within 20% difference), while it has been overestimated in section 3 surge at full speed test (+70%);
- number of surge cycles predictability can be assessed only in the ESD scenario. In this test, second and third HPRC sections coast down occurred in reverse flow condition consistently with the test run; even the timeframe relevant to the first section excursion in the Second Quadrant is matched exactly as per Figure 11. Anyway the dynamic simulation tended to overestimate the number of Surge cycles upon coast down (for section 1, simulation estimated 2 surge cycles instead of only 1), yielding however to a more conservative analysis of the ESD event.

6. Conclusions

As final step of the test campaign on HPRC, Surge exploration tests have been performed allowing an assessment of the machine design robustness and a data collection in reverse flow operation. The tests have been carried upon different scenarios of gas loop pressurization and control action on recycling valves, so to differentiate the Surge severity.

An important outcome is that reverse flow condition can be withstood by compressor even for a prolonged operation since vibrations and axial thrust trends decayed smoothly during the tests without any significant peaks amplitude. However, being Surge phenomenon associated to forward and backward gas flow, a threshold on the number of Surge cycles shall be taken into account to avoid undesirable upsets due to pressure waves and vibrations.

Data match between Surge explorations test and dynamic simulations, along with the characterization of the second quadrant branch, allowed to enhance and fine tune the capability of transient compressor modelling. A good level of consistency on the main parameters affecting the transient compressor operability has been demonstrated. With the aid of this activity, a more accurate evaluation of compressor Surge and reverse flow parameters (especially the number of Surge cycles) is achieved whenever a dynamic simulation is performed on a particular compressor application.

Surge exploration tests results analysis and dynamic modeling enhancement have been a breakthrough for the introduction of a new Surge Acceptance Criterion based on number of Surge cycles instead of time spent in the Second Quadrant. The new criterion is applicable not only to HPRC but also to standard compressors, with expected benefits in reducing complexity of Surge protection devices (like Hot Gas Bypass Valve or Cold Gas Bypass Valves) and relevant costs, still preserving safe operation of the compressor and of all the compressor system equipment.

Nomenclature

Variables
ω = Compressor speed
t = Time
T = Temperature, Torque
GR = Gear Ratio
f = Friction Loss factor of the train
I_{TOT} = Train total Inertia reduced to Compressor shaft
τ = Work coefficient
u = peripheral speed
D = Impeller diameter
φ = Flow coefficient
$\dot{m}$ = Mass flow rate
ρ = Mass density
r = Pressure ratio
P = Total Pressure
H = Head
η = Efficiency
Ψ = dimensionless polytropic head
Mu = Mach number

Subscripts
EM = Electric Motor
CC = Centrifugal Compressor
P = Process
Des = Design
pol = Polytropic
2 = Section at the external diameter of the impeller

Acronyms
HPRC = High Pressure Ratio Compressor
DE = Driver End
NDE = Non Driver End
ASV = Anti Surge Valve
ESD = Emergency Shut-Down
HGBV = Hot gas Bypass Valve
CGBV = Cold gas Bypass Valve
1xRev = Revolution Frequency

References

1. Falomi S., Iurisci G., Fattori S., Grimaldi A., Aringhieri C., Sassanelli G., Iannuzzi G. (2016), "Full scale validation of an high pressure ratio centrifugal compressor", *@ 45th Turbomachinery & 32nd Pump Symposia*
2. API Standard 617, "Axial and Centrifugal Compressors and Expander-compressors", 8th edition, September 2014
3. Belardini E., Tapinassi L., Rubino D.T., Pelella M. (2015), "Modeling of Pressure Dynamics During Surge and ESD", *@ 3rd Middle East Turbomachinery Symposium*
4. Belardini E., Rubino D.T., Tapinassi L., Pelella M. (2016), "Four Quadrant Centrifugal Compressor Performance", *@ Asia Turbomachinery & Pump Symposium*
5. Greitzer E.M. (1976), "Surge and Rotating Stall in Axial Flow Compressors. Part I: Theoretical Compression System Model", *Journal of Engineering for Power*
6. Greitzer E.M. (1976), "Surge and Rotating Stall in Axial Flow Compressors. Part II: Experimental Results and Comparison with Theory", *Journal of Engineering for Power*
7. Baldassarre L., Bernocchi A., Rizzo E., Fontana M., Maiuolo F. (2015), "Axial Thrust in High Pressure Centrifugal Compressors: Description of a Calculation Model Validated by Experimental Data from Full Load Test", *@ 44th Turbomachinery & 31st Pump Symposia*

481512 Input Design for Lti Systems with Multiple Time Scales

Abstract:

Input design for LTI Systems with Multiple Time Scales

Input design for multiscale systems is much more challenging than the single scale systems due to the presence of dynamics at multiple time scales. Conventional inputs such as PRBS are persistently exciting and are proven to be excellent inputs for the identification of LTI systems. In the case of systems with multiple scales, there is very weak overlapping between the bandwidths of fast and the slow subsystems unlike the single scale systems where there is significant overlap between the subsystems. This weak overlapping calls for additional identification friendly constraints on the input design unlike the single scale system. In order to excite the all the subsystems, the sampling has to be commensurate with the fast subsystem and the input should be designed such that it contains the fast subsystem bandwidth (which is relatively quite large compared to the slower one). When a non-sequential input such as PRBS is used for such systems, the slow subsystem is poorly excited as shown in Fig 1 unlike a single scale system which is shown in Fig 2. In Fig 1, we considered a system with time constants 5 and 0.05 sec while a system with time constant 5 sec is considered in Fig 2. The input bandwidth for the simulations is chosen based on the individual system bandwidths.

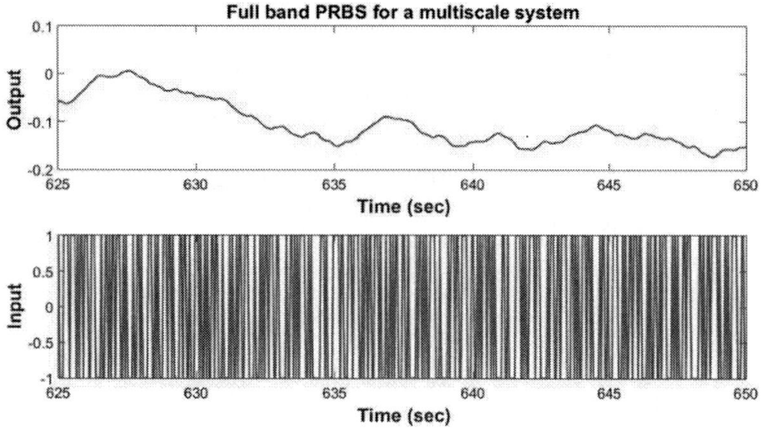

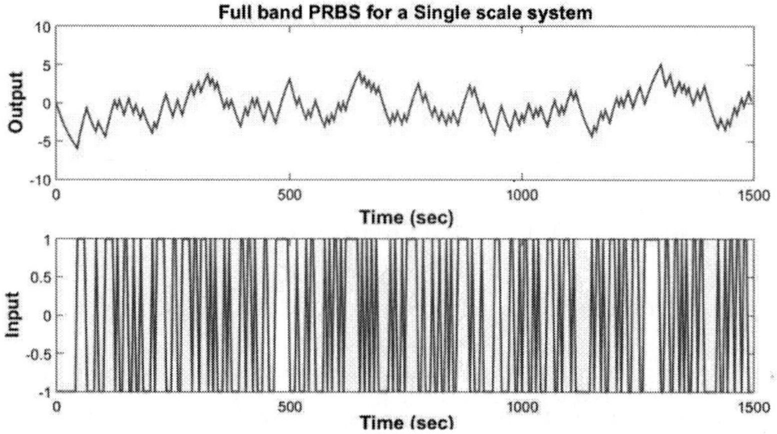

The reason for such poor excitation in the case of multiscale system is that the PRBS input excited all the frequencies in a non sequential manner and has sparsely given importance to the frequencies in the slow subsystem bandwidth. As a result, the observations obtained from such excitation resulted in a poorer information on slow subsystems. This can also be seen in an SNR perspective. In the case of single scale systems, including a input bandwidth outside the system bandwidth results in poorer SNR in the observations. A similar issue of a slightly different nature prevails in multiscale systems also. The contribution of fast subsystem behaves like noise while identifying the slow subsystem and vice versa along with the regular noise component. This is the reason for the poor SNR of slow subsystem as shown in Fig 1.

In this work, we propose an input that has non-sequential nature within a frequency band but a sequential nature across the frequency sub bands. The proposed input is more identification friendly for multiscale systems and the identifiability of the slow and fast subsystems is better than by using the conventional inputs.

480284 Solid Suspension in Stirred Vessels: Comparison of a Fully Coupled CFD-DEM and an Eulerian Multiphase Approach

Abstract:

Solid suspension in high viscous liquids and/or with high solids concentration is a challenging problem due to the solid-solid and solid-liquid phase interactions. Just suspended speed (N_{js}), suspension height (H_s) and the power consumption can vary not only with the material properties but also with the design of tank and impellers.

In the past, there was been work presented using multiphase CFD methods, specifically the Eulerian approach. This approach has shown promising results for turbulent conditions [1]. Further work has also been carried out where Discrete Element Modeling for particle flows was co-simulated with CFD simulations for fluid flow [2].

In this work we compare based on experimental data simulation results of fully 2-way coupled CFD-DEM simulations with an Eulerian Multiphase (EMP) approach based on the Kinetic Theory of Granular Flow (KTGF) in terms of accuracy and runtime. While the CFD-DEM simulation typically needs more runtime, it needs less model parameter determination and tuning than the EMP approach.

[1] R. Aglave, J. Feng, E. Janz, R. Strong and T. Eppinger, CFD Simulation of Solids Supension at High Loading, AIChE annual meeting 2015, Salt Lake City, UH, USA

[2] B. Blais, M. Lassaigne, C. Goniva, L. Fradette and F. Bertrand, 2015. A Robust and Efficient CFD-DEM Model for the Investigation of Viscous Solid-Liquid Mixing in Stirred Tanks, 10th European Conference on Chemical Engineering, Nice, France.

481111 A Refined Non-Random Two-Liquid Segment Activity Coefficient Model for Solubility Modeling

Abstract:

Accurate correlation and robust prediction for complex molecule solubility or solid-liquid equilibrium (SLE) in solvents and solvent mixtures have drawn much interest recently[1]. Solubility models such as the Non-Random Two Liquid Segment Activity Coefficient (NRTL-SAC) model[2] are now routinely used in the industry. This paper presents a refinement to the successful NRTL-SAC model to better account for the contribution due to the size differences of molecules. Specifically, the model is refined by substituting the Flory-Huggins combinatorial entropy expression with a modified Staverman-Guggenheim expression. Additionally, temperature dependency of activity coefficient is addressed in the refined model. We present the rationale for the refined model, along with the updated conceptual segment numbers for common solvents. Both the original and the refined NRTL-SAC models are employed to calculate vapor-liquid and liquid-liquid equilibrium for binary solvent systems and solubility of several drug molecules in solvents and solvent mixtures. The refined model provides improved prediction performance when compared with the original model, especially for systems containing larger size molecules. The refined NRTL-SAC model will be further extended to provide predictive capability for complex molecule solubility in excipients and polymers.

1. Tanveer, S.; Hao, Y.; Chen, C.-C., Introduction to Solid-Fluid Equilibrium Modeling. *Chemical Engineering Progress* **2014,** *110* (9), 37-47.

2. Chen, C.-C.; Song, Y., Solubility modeling with a nonrandom two-liquid segment activity coefficient model. *Industrial & engineering chemistry research* **2004,** *43* (26), 8354-8362.

481189 A Comprehensive Thermodynamic Model for the Lithium Chloride and Water Binary System Using the Electrolyte NRTL Model

Abstract:

Thermodynamic properties of the lithium chloride and water binary system have been extensively studied, however, many of these analyses are applicable only to a narrow range of concentration and/or temperatures. The objective of the present work is to establish a comprehensive thermodynamic representation by implementing the symmetric electrolyte NRTL (eNRTL) activity coefficient model. The solubility model correlates the nonideality of the solution with the composition dependency by utilizing two binary interaction parameters for each of the electrolyte-electrolyte and molecule-electrolyte pairs in the system. Each parameter is determined by applying up to three temperature dependent coefficients and then regressed from experimental data. This allows the parameters to establish the accuracy of the model for concentrations up to saturation and temperatures between 273.15-473.15 K. The advancement of process design, optimization, and operation range of absorption heat pumps and refrigeration systems, as well as the increasing interest in understanding the phase behavior of lithium salt systems for energy storage devices facilitates the need for a comprehensive thermodynamic model.

481495 Thermodynamic Modeling of HCl-H_2so_4-H_2o Ternary System with Electrolyte NRTL Model

Abstract:

Abstract

The hydrochloric acid-sulfuric acid-nitric acid system plays a critical role in stratospheric aerosol formation which leads to ozone layer deposition. Recently, this mixed acid system has gained a lot of attention in chemical nitration processes for energetic materials and in recovering noble metals from electronic waste. It is therefore important to develop for the mixed acid system a comprehensive thermodynamic model which could accurately represent various thermodynamic properties such as vapor pressure, molar enthalpy, molar heat capacity, and salt solubility. Our previous work [1] reported a comprehensive thermodynamic model for the HNO_3-H_2SO_4-H_2O ternary system based on the electrolyte NRTL model [2]. In this work, we extend this work for the HCl-H_2SO_4-H_2O ternary system with a temperature range of 273 to 373 K and H_2SO_4 concentration from 0 to 100 wt%. The model should be very useful in supporting heat and mass balance calculations and process simulation for processes involving HCl and H_2SO_4.

Keywords:

Hydrochloric acid, Sulfuric acid, Electrolyte NRTL model

References:

[1] M. Wang, H. Kaur, C.-C. Chen, "Thermodynamic Representation of Nitric Acid-Sulfuric Acid-Water Ternary System," manuscript submitted to AIChE Journal (2016)

[2] Y. Song, C.-C. Chen, "Symmetric Electrolyte Nonrandom Two-Liquid Activity Coefficient Model," Industrial & Engineering Chemistry Research, 48 (2009) 7788-7797

480288 Calculation of Optical Properties of Molecular Dyes Using Time-Dependent Density Functional Theory (TD-DFT) and Development of Quantitative Structure-Property Relationships (QSPR)

Abstract:

Optical polarizers play an important role in liquid crystal display (LCD) and organic light emitting diode (OLED) technologies. There has been an increasing demand for new polarizing films with innovative structure and reduced film thickness (e.g., thin coatable polarizers) for applications in mobile devices, high-end TVs, gaming, among others. High performance dye molecules with large dichroic ratios have been long sought after as they are key for achieving high polarization efficiencies in a coatable polarizer. Other optical properties, such as birefringence, are also crucial for the performance of polarizers in LCD and OLED devices. Birefringent materials are currently used to modify the polarization state of light passing through it, which allows observing different contrasts depending on viewing angle.

In this presentation, TD-DFT methods will be proposed as a valuable tool for the accurate calculation of optical properties of molecular dyes. The methodology used for the calculation of birefringence will be explained in detail and the methodology used to develop a QSPR for dichroic ratios of dyes as a function of molecular descriptors will also be presented.

483581 Opportunities and Challenges for Process Innovation and Intensification: An Industrial Perspective

Abstract:

Despite more than a century of developments in chemical processing, large opportunities for energy and atom efficiency remain. Increasing population means that demand for products of the industry will continue to grow, posing a challenge to shrink the environmental footprint per unit of production.

This talk will examine some of the process improvement opportunities of interest to Dow and how we have attempted to address them. In addition to highlighting some success stories, the talk will present some of the barriers we encounter when attempting to innovate and intensify process technologies.

483582 Intensification of Processes yielding Renewable Bio-Products: Theory - Experiments - Field Applications

Abstract:

The historical journey of Process Intensification is first reviewed, and placed in context of prior research activities. Then the potential role of Process Intensification and the AICHE RAPID Institute in the development of novel processes for the creation of Renewable Bio-Products will be outlined. Emphasis will be given to a wide array of theory tools that can be brought to bear for the synthesis of novel processes yielding renewable bio-products, experiments that can be used for the validation of these theory tools, and the practical development of proposed new technologies, and a number of field applications that can serve as demonstration vehicles for the proposed technologies.

479900 Energy Innovation: Challenges to the Commercialization of New Energy Technology

Abstract:

Meeting growing global demand for energy in the 21st century in an environmentally responsible manner will require a significant escalation in innovation leading to the development of new, clean energy sources. Innovation is defined as invention plus commercialization, so that the innovation process isn't complete until the invention has been turned into something that provides real world benefits. Innovation in energy is complex, has a high rate of failure and takes a long time. Commercializing new energy technology requires numerous technical and business disciplines, partnerships with large and small public and private entities, support from service organizations and most importantly, significant investment.

This presentation will explore the energy innovation process, the role of an innovation ecosystem and the benefit to industry of supporting early stage technology companies. It will also look at how business accelerators and incubators can reduce the risk of failure and speed up the innovation process by helping energy entrepreneurs navigate the commercialization process, develop support networks, acquire skills and resources, develop strategic relationships and attract funding.

480267 The Right Combination - New Products from Readily Available Feedstocks

Abstract:

The global commodity chemicals market is in a state of subtle but constant evolution. Understanding the need and supplying the right product at the right price is often a challenge. Even when the process involves known chemistry, feed stock cost and availability can be limiting, particularly when trying to enter a high volume/low margin market. A concept whereby new products may be manufactured from readily available feed stocks using known chemistry is presented. Topics such as targeting the right molecule, process design and scale-up, product testing, and process safety will be discussed.

Imperial College London

UNIHEAT

Engineering Innovation by Design: The Uniheat Project

Sandro Macchietto[1,2] and Francesco Coletti[2]

[1]Imperial College London, [2]Hexxcell Ltd

2017 AIChE Spring Meeting – Session: Technology Transfer: The process and the Applications
San Antonio, Tx, USA - March 26-30, 2017

Innovation is hard

UNIHEAT

US Venture capital funds 1998-2000

→Google, Amazon

$175bn invested – mostly wasted

VC funds 2000-2004

Average rate of return less than 2%

There is no silver bullet
Can we change the odds?

2

The process of innovation

UNIHEAT

Traditional, linear approach

Science

↓

Engineering

↓

Applications

↓

Products/Services

↓

$

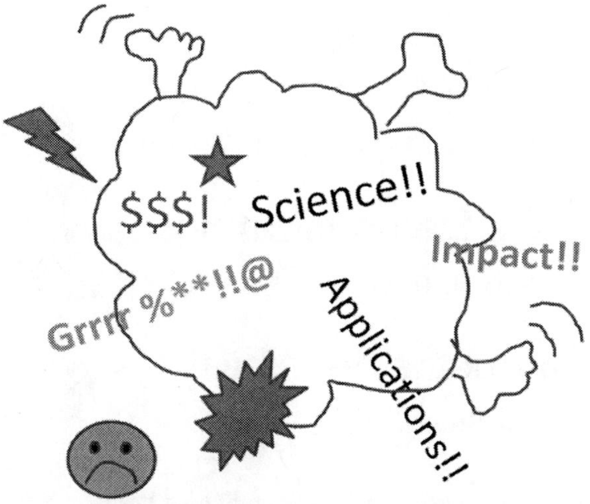

Innovation by accident

3

The process must be "engineered"

UNIHEAT

Traditional, linear approach

Science

↓

Engineering

↓

Applications

↓

Products/Services

↓

$

Structured Approach + Experience

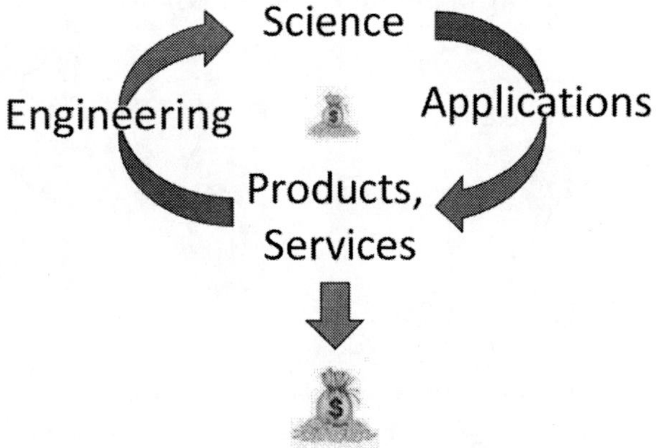

Innovation by design

4

3 Steps to Innovation – with industry

UNIHEAT

UNIHEAT	Industry partners
• **Advanced Research** methods, data, software	**Engagement** focus, trust, relevance
• **Lab Demonstration** small/medium scale	**Data, materials** proof of concept
• **Industrial feasibility** full scale, real	**Plant, people** performance, benefit, ... steps for use

Greater probability of Technology Transfer success

The Problem

UNIHEAT

Oil Refineries use 5-7% of crude Oil to operate - World-wide >5m bbl/day !!

(Saudi Arabia production ~10m bbl/day)

2.5% of all man-made CO2 emissions

Key refinery processes

Reaction and conversion

 wrong products, high temperatures

Thermal energy management

 fouling, poor energy integration

Refinery design and practice

Support Service practice

Macchietto et al.,
Fouling in crude oil preheat trains: a systematic solution to an old problem", HTE (2011)

6

The Opportunity

UNIHEAT

Overall, over 25% of refinery **energy losses** could be practically recovered.

In **some** process units **more than 50%** (US DOE 2006)

current average energy used

practical minimum energy

theoretical minimum energy required

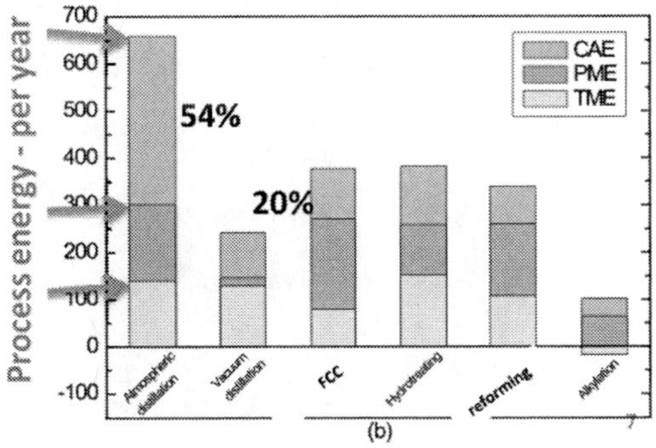

Fouling in Refinery CDUs

UNIHEAT

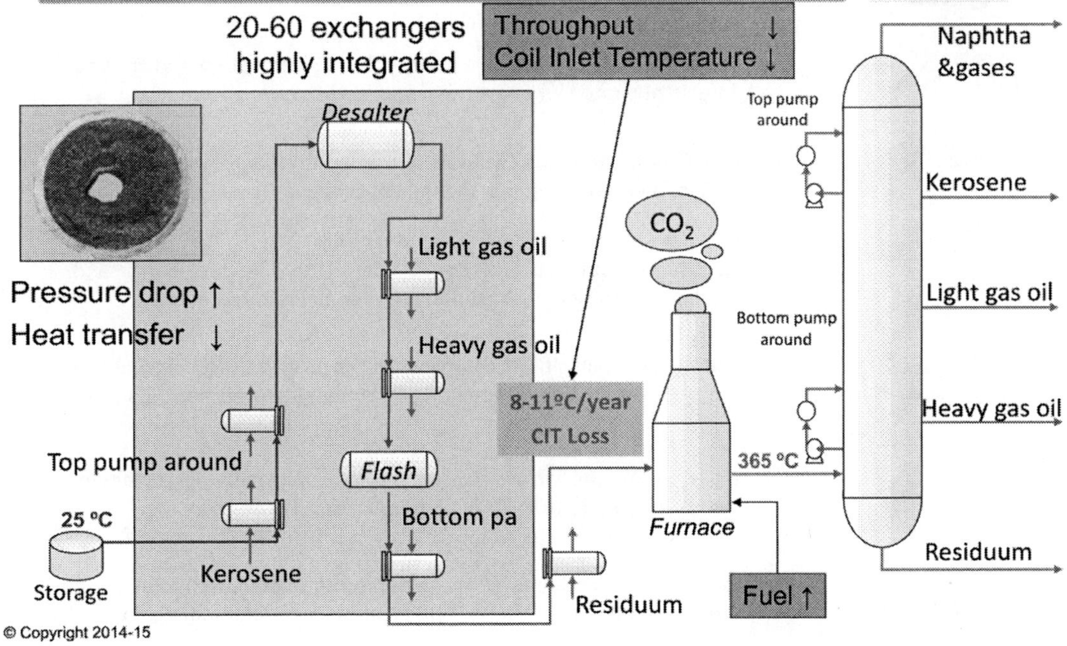

The UNIHEAT Project

UNIHEAT

- **An international partnership between leading research institutions and industry**
- **A well funded applied research programme with industrial engagement built-in**
- **The largest interdisciplinary team anywhere in the world in this area in recent years**

www.uniheat-project.com

Project partners

www.uniheat-project.com

UNIHEAT

Imperial College London	**Imperial College London** is a science & engineering based institution consistently rated amongst the world's best universities.
ИК СО РАН	**Boreskov Institute of Catalysis** is the world's largest research and development institution in the field of catalysis.
Novosibirsk State University НГУ	**Novosibirsk State University** is one of the best research and teaching institution organisation based in Novosibirsk, Russia.
UNICAT	**UNICAT Ltd** is a spin-off company that focuses on R&D activities in the fields of catalysis, engineering and energy efficient technologies.
bp	**BP** is one of the world's largest oil companies serving customers every day in more than 90 countries across six continents.
Sk Skolkovo	**Skolkovo Foundation** is a Russian non-profit organization founded in 2010 to create a new science and technology development centre

10

4 Components

UNIHEAT

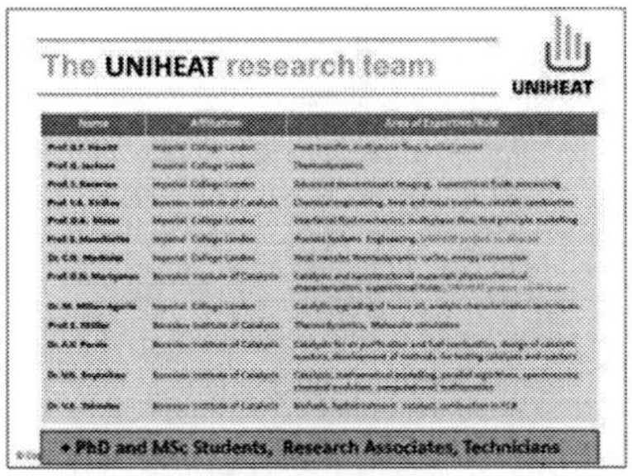

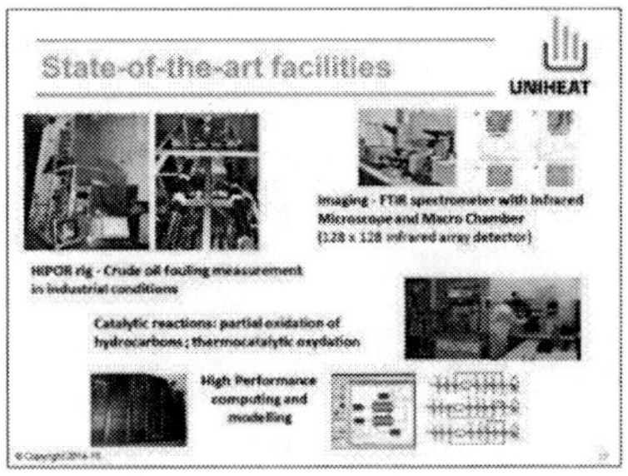

11

The UNIHEAT research team

UNIHEAT

Name	Affiliation	Area of Expertise/Role
Prof. G.F. Hewitt	Imperial College London	Heat transfer, multiphase flow, nuclear power
Prof. G. Jackson	Imperial College London	Thermodynamics
Prof. S. Kazarian	Imperial College London	Advanced spectroscopic imaging, supercritical fluids processing
Prof. V.A. Kirillov	Boreskov Institute of Catalysis	Chemical engineering, heat and mass transfer, catalytic combustion
Prof. O.K. Matar	Imperial College London	Interfacial fluid mechanics, multiphase flow, first principle modelling
Prof. S. Macchietto	Imperial College London	Process Systems Engineering, UNIHEAT project co-director
Dr. C.N. Markides	Imperial College London	Heat transfer, thermodynamic cycles, energy conversion
Prof. O.N. Martyanov	Boreskov Institute of Catalysis	Catalysts and nanostructured materials physicochemical characterization, supercritical fluids , UNIHEAT project co-director
Dr. M. Millan-Agorio	Imperial College London	Catalytic upgrading of heavy oil, analytic characterization techniques
Prof. E. Müller	Boreskov Institute of Catalysis	Thermodynamics, Molecular simulation
Dr. A.V. Porsin	Boreskov Institute of Catalysis	Catalysts for air purification and fuel combustion, design of catalytic reactors, development of methods for testing catalysts and reactors
Dr. V.N. Snytnikov	Boreskov Institute of Catalysis	Catalysis, mathematical modelling, parallel algorithms, spectroscopy, chemical evolution, computational mathematics
Dr. V.A. Yakovlev	Boreskov Institute of Catalysis	Biofuels, hydrotreatment catalyst, combustion in FCB

+ PHd and MSc Students, Research Associates, Technicians

12

State-of-the-art facilities

UNIHEAT

HIPOR rig - Crude oil fouling measurement in industrial conditions

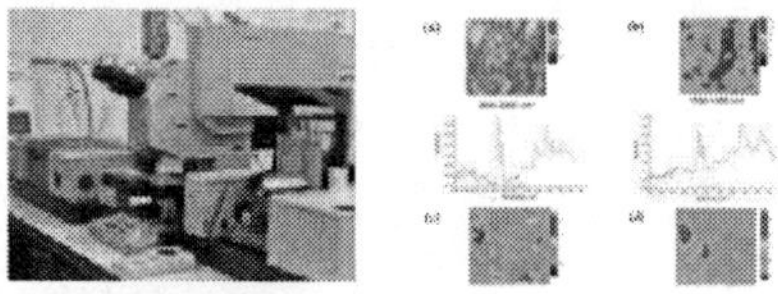

Imaging - FTIR spectrometer with Infrared Microscope and Macro Chamber
(128 x 128 infrared array detector)

Catalytic reactions: partial oxidation of hydrocarbons ; thermocatalytic oxydation

High Performance computing and modelling

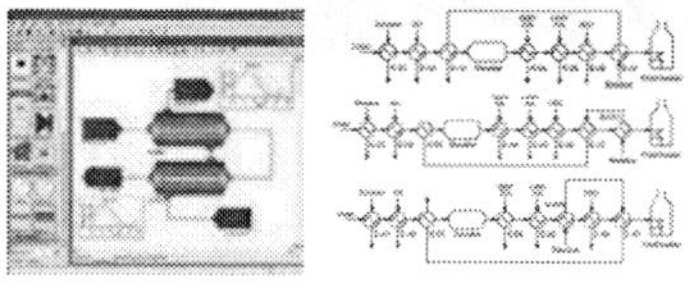

13

From molecular to plant scale

UNIHEAT

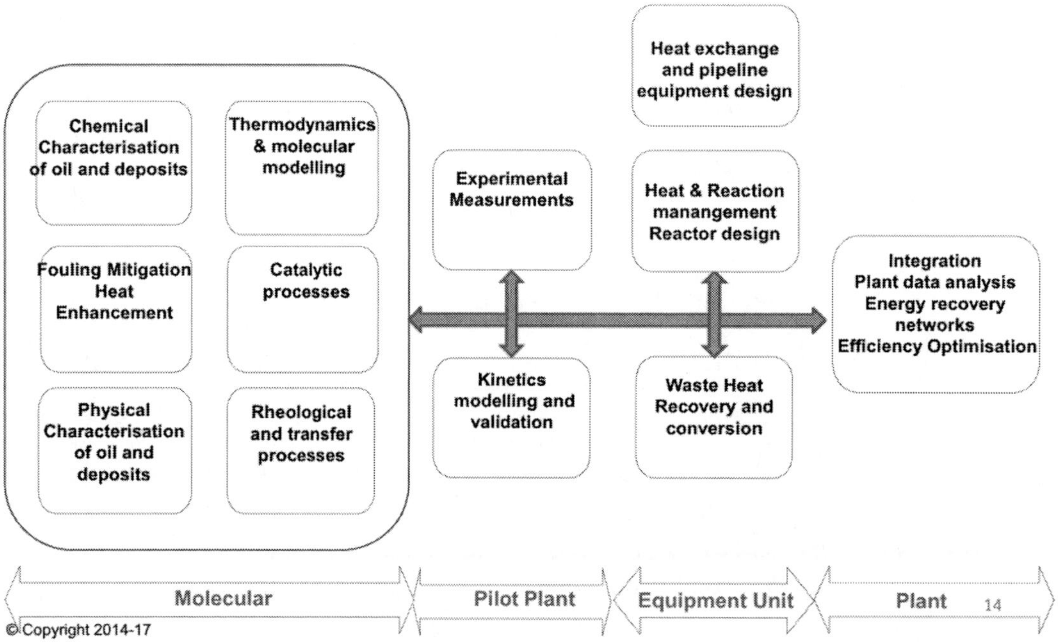

14

6 Research Themes

UNIHEAT

15

Research students, workshops, exchanges, …

UNIHEAT

Theme 1 – Multi-scale modelling of crude oil fouling

UNIHEAT

Thermo & Molecular modelling
Prof. E. Müller and Prof. G. Jackson

- Experimental data
- Equation of state → Properties of bulk fluids
- Parameters intermolecular potential
- Molecular simulation → Interfacial & transport properties

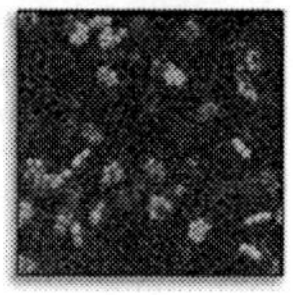

Fundamental transport phenomena
Prof. O. K. Matar

- Heat balance → Heat transfer coefficient
- Momentum balance → Shear stress & pressure drops
- Chemical reaction & equilibrium → Fouling deposition & removal

Industrial scale modelling
Prof. S. Macchietto

- Shell-side coupled to multi-pass tube-side model
- Costing and impact model → Improved operations / Optimal design / Economical retrofit

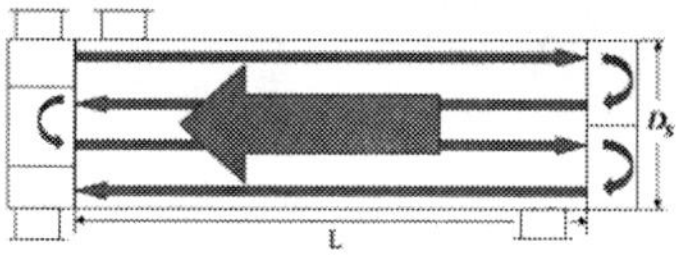

Image courtesy of Hexxcell ltd.

CDU Pre-heat train: Refinery X
Thermo Hydraulic-based Cleaning scheduling

UNIHEAT

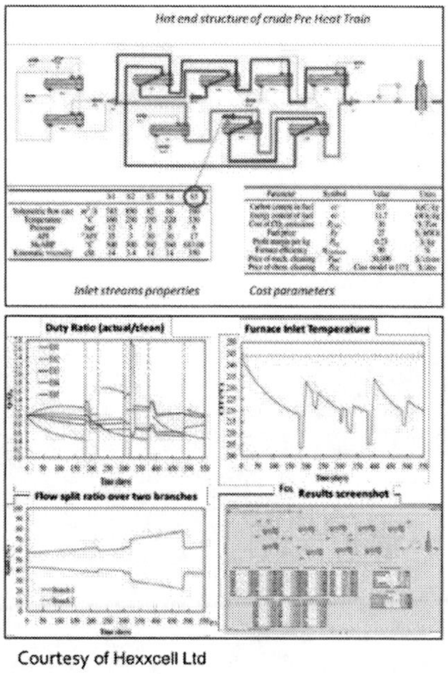

Courtesy of Hexxcell Ltd

Some issues

- Back pressure controlled
- Flow balancing (split) in two branches
- Complex interactions – e.g. Stream 5
- Cleaning 1 exchanger shifts flows, fouling, Duty and Pressure drops elsewhere
- Thermal (TL) and hydraulic (HL) limits

Results enable accounting for and assessing:

- Hydraulic as well as thermal effects and limits
- Effect of cleaning an HEX on shift of fouling rates, heat duty and pressure drop to other HEXs in overall network
- Depth and effectiveness of a chemical cleaning, conditions an HEX is left in, and impact on later fouling build-up as function of operation
- "Condition-based" cleaning to maximises chemical cleaning effectiveness
- Technical and economic performance of alternative cleaning types (chemical and mechanical) and cleaning schedules (thermal + hydraulic impact, costs of energy & cleaning, refinery margins benefit)

➔ More reliable predictions, more realistic schedules better decision

L. Lanchas-Fuentes et al. Management of cleaning types and schedules in refinery heat exchangers, Proceedings, HEFAT2016, Malaga, Spain July 11-13 2016

Theme 2 – Crude oil and fouling deposit characterisation

UNIHEAT

ATR-FTIR Spectroscopic Imaging

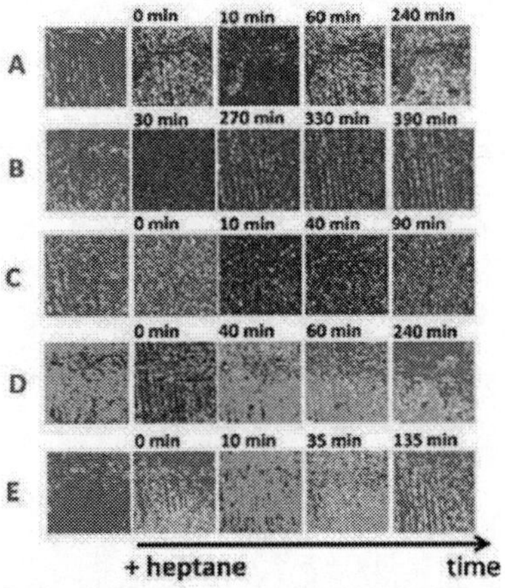

- range of advanced spectroscopic and chemical imaging methods (ATR-FTIR, NMR, ESR and SAXS on same oils
- complementary information at different spatial and temporal scales
- *First In-situ* chemical imaging analysis of asphaltenes aggregation, instability, deposition

Oil characterization techniques - **Joint work: ICL + BIC + BP**

Gabrienko, Subramani, Martyanov and Kazarian. Correlation between Asphaltene Stability in n-Heptane and Crude Oil Composition Revealed with In Situ Chemical Imaging Adsorpt. Sci. Technol. 2014, **32**, 243–255.

Asphaltene stability

UNIHEAT

FTIR Imaging

NMR Imaging

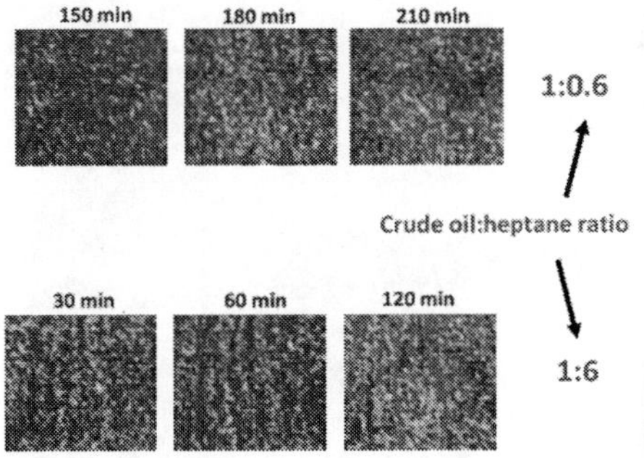

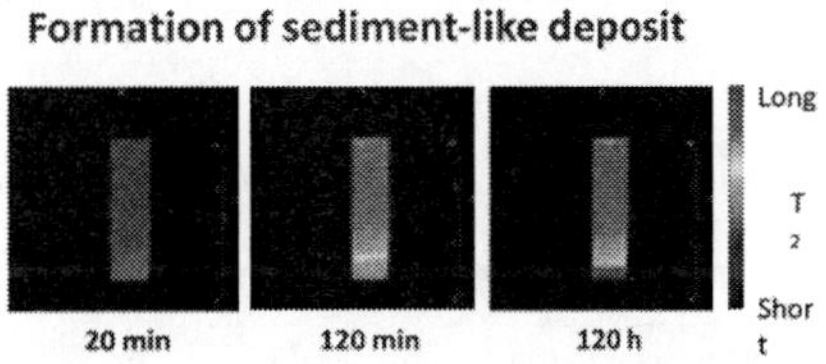

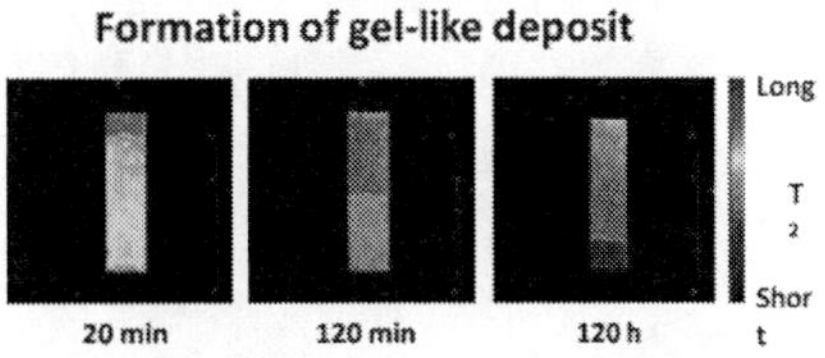

Gabrienko A.A., et al. J Phys Chem C, 119, 2015, 2646-2660.

20

Credits: Prof. S.G. Kazarian and Prof. O.N. Martyanov

Molecular modelling

UNIHEAT

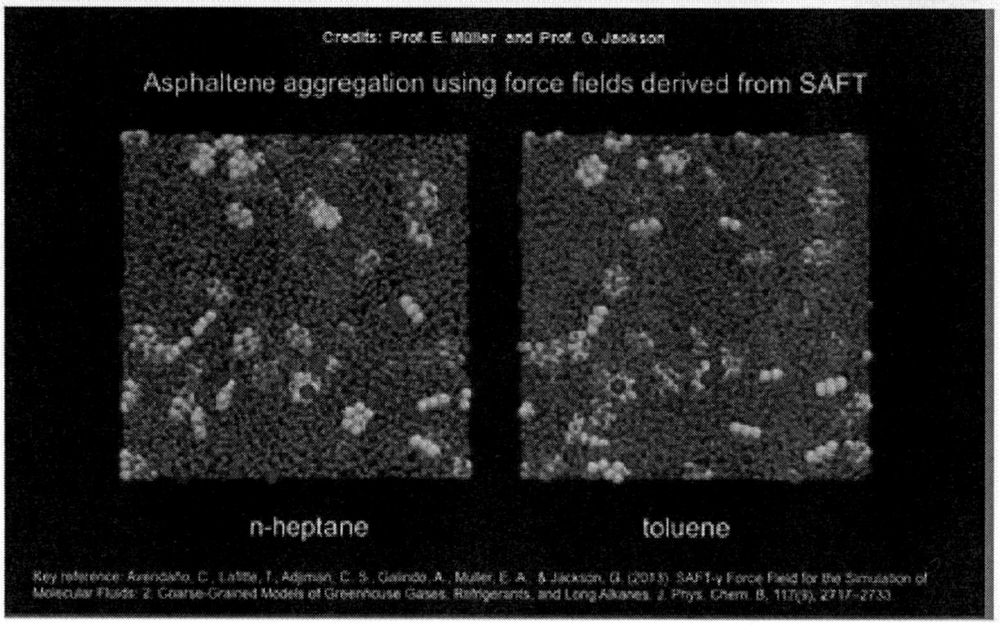

Theme 2 - Flameless industrial heaters

Drag reduction in pipelines

UNIHEAT

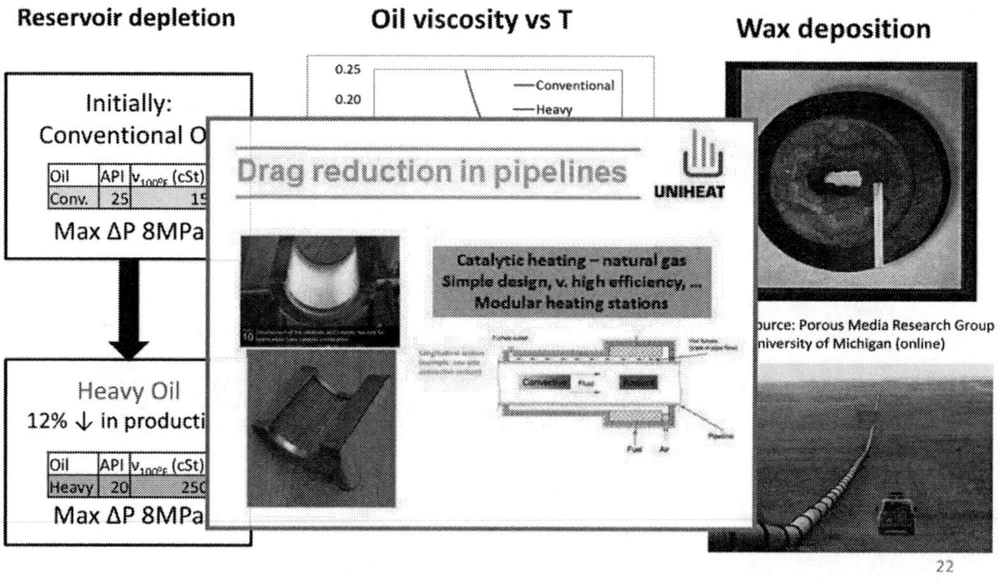

Flameless catalytic heating
for all kind of industrial heaters

UNIHEAT

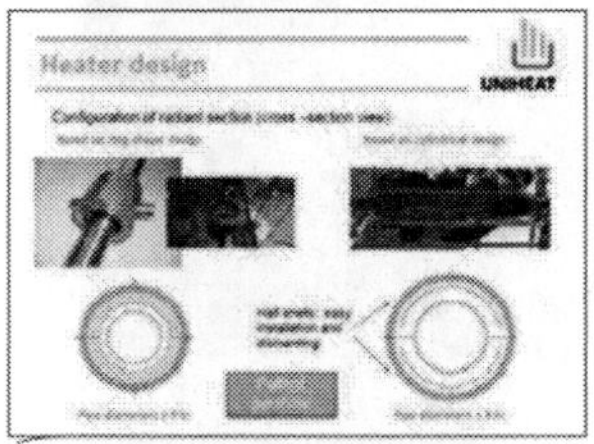

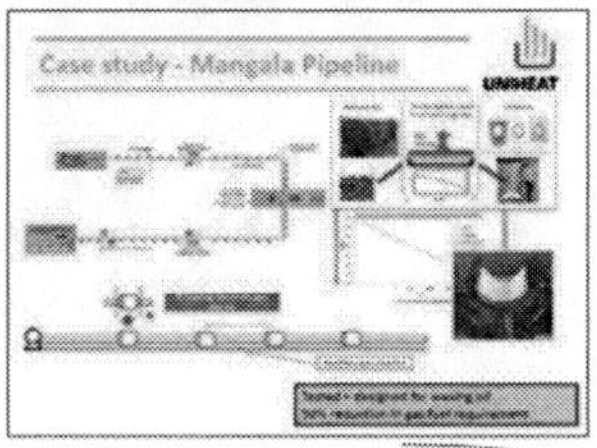

Full scale ?

Portfolio of patents, technology, knowhow

Key aspects

UNIHEAT

Features

- Flameless combustion
- Air and fuel fed at different locations prevent explosive mixtures

Benefits: Improved safety

- Very high conversion
- Low NO_x emissions

Benefits: Fuel & environmental efficiency

- Precise control over a wide range of power outputs
- Variety of gas fuels
- Variety of geometries for:
  - Improved gas distribution
  - Heat flux directional control
  - Efficiency
- Advanced flameless start-up

Benefits: Flexible design and operation

24

Theme 4 - Thermodynamic power generation cycles for improved efficiency

UNIHEAT

Low grade waste-heat recovery

Catalytic heater (BIC)

Organic Rankine Cycle (Imperial College)

ORCat – Concept : High-efficiency, low-emission, fuel-flexible Combined Heat and Power (CHP) system

25

Low grade waste-heat recovery

UNIHEAT

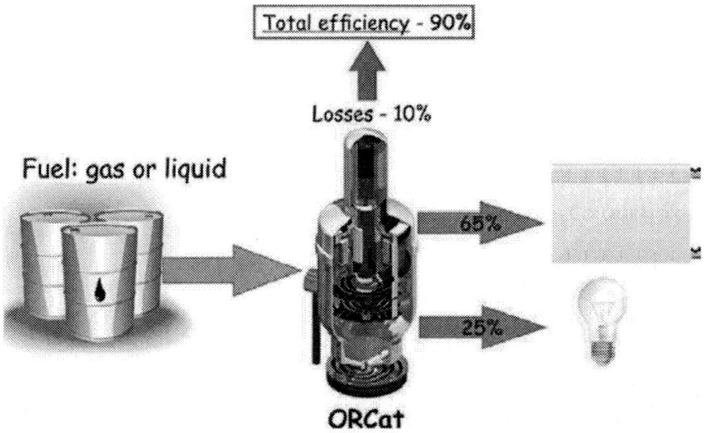

ORCat – Concept : High-efficiency, low-emission, fuel-flexible Combined Heat and Power (CHP) system

26

Industry Engagement Programme

UNIHEAT

Imperial College London

R&D

Industry partners

Spin-offs, SMEs ...

Implementers and End-users

IMPERIAL CONSULTANTS

Engagement and knowledge transfer

Imperial Innovations

Technology transfer

Tech transfer workshops

UNIHEAT

Novosibirsk - 24 April 2014

Technology audits

UNIHEAT

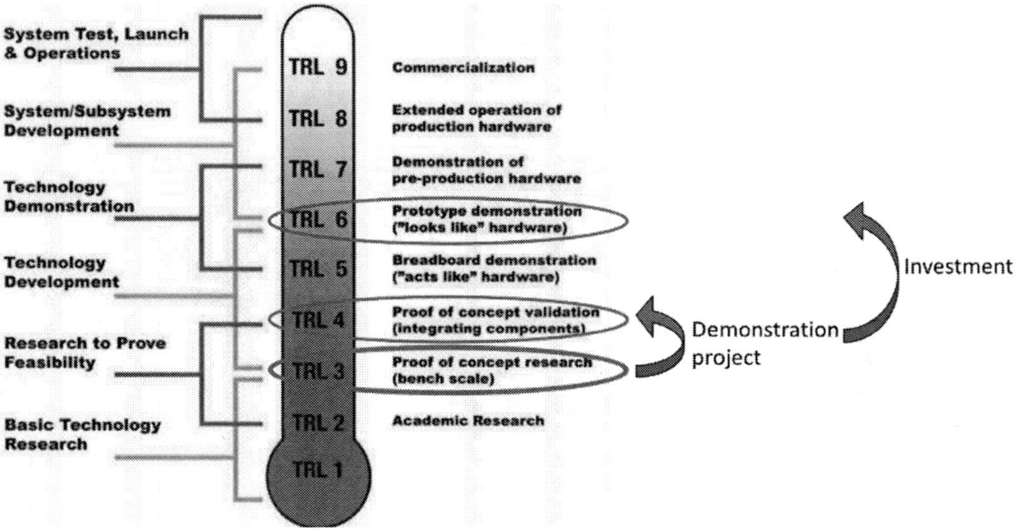

29

Industrial visits and exhibitions

UNIHEAT

Events

Europe, Russia, USA, Middle East

UNIHEAT

Partnerships

UNIHEAT

Houston Technology Centre

Key Outcomes: Technologies

UNIHEAT

- Fouling in oil refineries (measurement, modelling and mitigation)
- Crude oil characterisation
- Catalytic heating for pipelines and industrial heaters
- Catalytic combustion of heavy oils
- Heat recovery via catalytic ORC engine
- Laser induced pyrolysis of light hydrocarbons

33

Key outcomes (by end of 3-yr project)

UNIHEAT

- Research
  - Novel approaches, ideas, data
  - Over 50 journal papers, >60 conf.
- Industry engagement
  - Technology demonstrations
  - 22+ patents
  - 3 spinoffs
  - New contacts & collaborations
- Training
  - >60 researchers
- Long term collaboration in specific areas

22 patents filed, demonstrators, 3 spinoffs ...

Efficient R&D ?

UNIHEAT

Technology transfer

	MIT	IMPERIAL	UNIHEAT		Achieved
Turnover	per annum	per annum	per annum Reasonably expected	TOTAL	
Research £	440m	300m	3 m		
Staff	10,000	6,000	12		
Disclosures	632	350	3.5(£) – 0.7(staff)	→ 5-10?	
Patents	153 (20%)	50(15%)	<1	→3-5?	22+
Spinoffs	16-26	3-6	~0.2	→ 1-2?	3
Lead time		2-3 years		year 3?	

Инвестиционный Меморандум
КОНФИДЕНЦИАЛЬНО

1

35

Awards

UNIHEAT

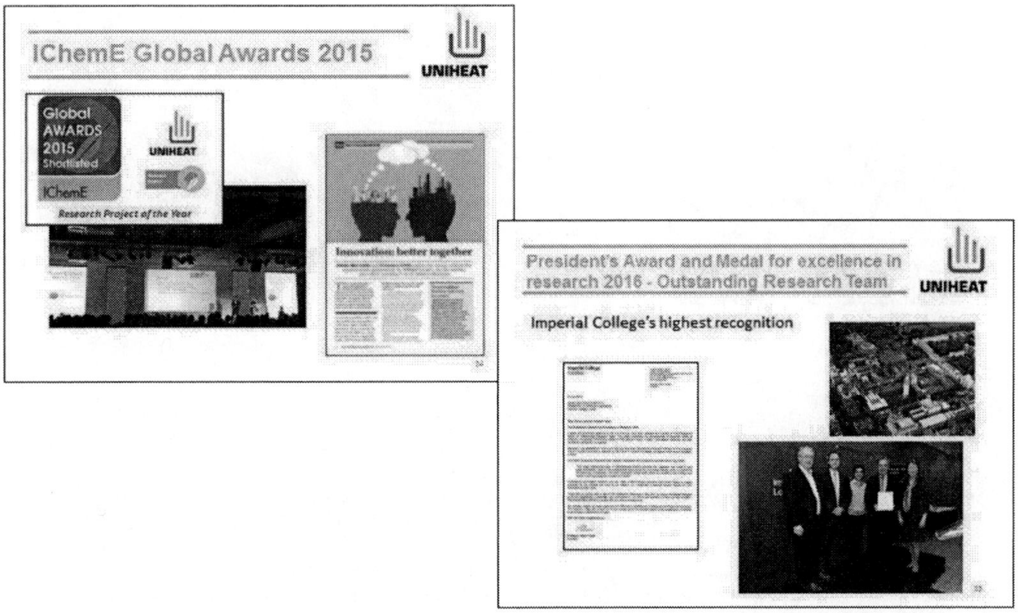

No single bullet …

Fast track research, knowledge transfer, tech transfer, …

… we can change the odds !

Macchietto Sandro and Francesco Coletti (2015). Innovation: better together.
The Chemical Engineer, 893, 24-27. (**Cover article**)

37

…it is a people business !

Innovation: better together

Thanks to

Participating Institutions, Sponsors, Companies
Colleagues, Researchers, Students, Technicians, Support staff
(many others, … UKTI, Embassies, …)

Thanks

Any questions?

s.macchietto@imperial.ac.uk
f.coletti@hexxcell.com

39

480303 Advanced Simulations in Pilot Plant Process Development for Hydrotreating Bioderived Feedstocks to Finished Specification Fuels

Abstract:

Process simulations have been a staple tool in process development work. Advanced simulations offer the opportunity to evaluate process performance risk for process development at low cost, if utilized correctly. As the government continues to push for lower emission transportation fuels, hydrotreating various alternative feedstocks (fisher tropes liquids from biomass gasification and plant and animal oils) to finished fuels continues to be the most conventional way to produce lower emission transportation fuels. Integrating new bioderived feedstocks into conventional hydrotreating is ever becoming more important. Southwest Research Institute (SwRI) has a long history of conducting pilot-scale hydrotreating, historically the expertise is gained from empirical data. As we transition to lower cost modelling efforts and better understanding the impact of bioderived feedstocks on conventional hydrotreating, advanced simulation options are being tested against historical pilot plant data and existing spreadsheet models. The presented case study evaluates advanced simulation capabilities for evaluation of alternative feedstocks in conventional refinery hydrotreaters by comparing empirical pilot-scale data to model performance.

AUTHOR INDEX

AUTHOR INDEX